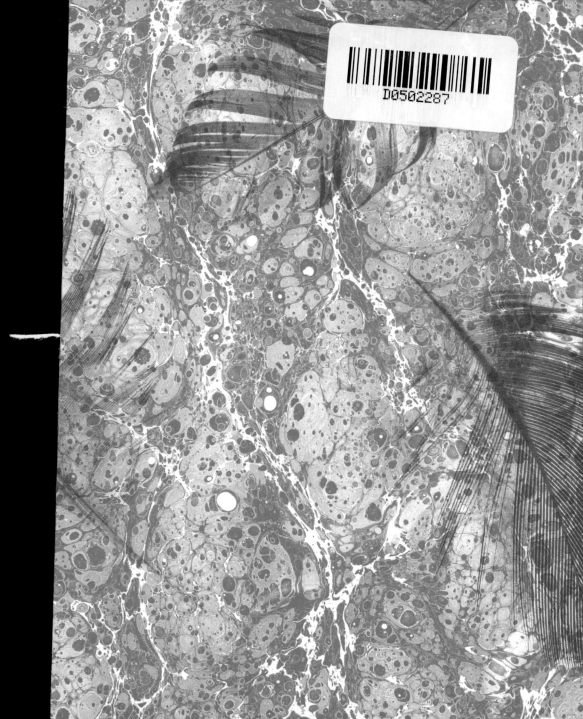

Pasta *for*
NIGHTINGALES

Pasta *for* NIGHTINGALES

A 17TH-CENTURY HANDBOOK *of* BIRD-CARE *and* FOLKLORE

Giovanni Pietro Olina

*With Illustrations from the "Paper Museum"
of Cassiano dal Pozzo*

Yale

UNIVERSITY PRESS

New Haven and London

NOTE FOR THE READER

Pasta for Nightingales brings together text from an Italian ornithological treatise of 1622 with the original coloured drawings of birds that served as the basis for its illustrations. The drawings are a selection of those created for the Paper Museum, so-called, of Cassiano dal Pozzo, which are now preserved in the Royal Library at Windsor Castle. Cassiano dal Pozzo (1588–1657) was one of seventeenth-century Italy's most prolific and untiring patrons of the arts and sciences, and his Paper Museum was an extraordinarily bold attempt to record any and every subject that drew his interest. These included many aspects of natural history, including the present bird drawings by the artist Vincenzo Leonardi (1589/90–1646). Leonardi was responsible for all but two of the studies of birds included here.

The text is drawn from Giovanni Pietro Olina's *Uccelliera* (or to give it its full title, *L'Uccelliera, ovvero discorso della natura e proprietà di diversi uccelli*, Rome 1622), excerpts from which have been translated here for the first time. The excerpts are paired here with the relevant drawings specially commissioned by Olina's patron, Cassiano dal Pozzo, to illustrate the treatise.

The *Uccelliera* was partly based on an earlier ornithological treatise by Antonio Valli da Todi, *Il canto degl'augelli* (Rome 1601), from which several of the illustrations were carried over. These included many of the prints of hunting and bird-catching scenes by other artists, some of which are also reproduced here.

The translation aims to provide a feel for the seventeenth-century vernacular used in the original, including the bird names,

which are thus not always the ones that would be associated by modern ornithologists with the particular species described (see, for example the differing identifications given on pages 86 and 87). Modern identifications are therefore provided in the captions to the watercolours, while the reader will additionally find scientific names for the birds depicted given alongside the common names in the index. These are taken from the Royal Collection's catalogue of the complete corpus of ornithological drawings by Leonardi and the other artists working for Cassiano (Part B.IV–V of *The Paper Museum of Cassiano dal Pozzo: A Catalogue Raisonné*, 2017), where the identifications were provided by ornithologists Carlo Violani and Edward Dickinson.

It should be noted that the drawings are not reproduced at actual size or to scale, that the orientation has occasionally been changed, and that – as cut-outs – all of the drawings have been trimmed, including the seventeenth-century inscriptions giving the birds' names in Italian and Latin. Dimensions are therefore not provided. The drawings vary in size, but do not as a rule exceed 160 x 200mm, having already been trimmed after they entered George III's collection in the eighteenth century. It was at this time, during the process of laying the drawings on new mounts, that many in the group suffered the water damage that is still visible in several of the illustrations here. In every case the engraved image in the *Uccelliera* was reversed from the original drawing, and, as Olina was keen to point out, most of the plates were life-size. The drawings are available online, and can be searched by Royal Collection Inventory Number (RCIN): www.royalcollection.org.uk/collection.

Contents

Foreword

HELEN MACDONALD

MAGINE YOURSELF IN an artist's studio in seventeenth-century Rome. A man called Vincenzo Leonardi frowns in concentration as he lays thin washes of watercolour over a drawing in black chalk. In front of him is the bird he is painting, a rock sparrow. Like all the birds he has been commissioned to draw for this project, his model is real. Perhaps it is dead, an artfully arranged corpse bought from a market bird-catcher. Or maybe it is a live-trapped specimen placed in a cage for his perusal. It could even be a hand-reared sparrow tame enough to preen as he paints all its points. There is no way we can ever know the true history of this creature. But as its feathers are drawn in, and as Leonardi elaborates upon them with lead white, and eventually deepens them with a glaze of gum arabic, the astonishing vivacity of the real bird is miraculously recreated. From the drawing, one might think the sparrow could hop from the page, chirrup, fly back to the mountains from which it came.

Pasta for Nightingales is the first English translation of a 1622 treatise in Italian called the *Uccelliera*, or 'The Aviary', written by Giovanni Pietro Olina and subtitled '[A] Discourse on the Nature and Distinctive Characteristics of Diverse Birds, and in particular of those which sing, together with the way of catching them, recognising them, raising them and maintaining them.'[1] Illustrated by careful engravings largely based on Leonardi's paintings, the book is no mere birdkeepers' almanac. The man who commissioned it, the scholar and patron of art Cassiano dal Pozzo, used it as his entry presentation to join the Accademia dei Lincei (Academy of the Lynx-eyed) a select group of men who sought to discover the truths of nature through keen and diligent observation. Commissioning realistic paintings of specimens from skilful artists was part of their strategy, and over his life-time Cassiano assembled an encyclopedic collection. He called it his *Museo Cartaceo*, or 'Paper Museum'. A vast visual catalogue of everything from antiquities to curious fungi, misshapen citrus fruits, fossils, sea fish, mammals and birds, it became a renowned resource for scholarly research and discussion. A large part of it was acquired by George III in 1762 and is now part of the Royal Collection, including the drawings of birds from which the images in this book are taken.

Modern scholars view Olina, who worked in Cassiano's household, more as the collator of the *Uccelliera* than as its sole author. Some of the book was based on an earlier work by Antonio Valli da Todi, but it is likely that much of it was written by Cassiano himself. He and his younger brother were keen and knowledgeable bird enthusiasts, keeping species as varied as flamingoes, bearded vultures, herons and parrots at their house in Rome.

The *Uccelliera* is many things. One is a product of the scientific revolution: a book that rejects many previously unquestioned notions about birds, setting store instead on the 'exact knowledge of Nature that might be obtained by long observation and contemplation'. The founder of the Accademia dei Lincei, Federico Cesi, considered its illustrations of birds marvellous expressions of the subjects represented, like nothing else he

Title page of the *Uccelliera*

had seen. He described the *Uccelliera* as 'without a doubt of great benefit to natural history, the true base and foundation of good and open philosophy'.[2]

But this treatise is also a fascinating window onto a long-lost world. Today most of us think of birds as creatures simply to be watched in the wild, and consider birdkeeping far removed from scientific ornithology. The role of birds in seventeenth-century Italy can feel bewildering to us: so many species were simultaneously viewed as delightful songsters, culinary delicacies, and useful human medicine. It was a time where gorse blooms were fed to captive robins to keep them merry; where lapwings were kept in gardens for their beauty and utility in pest control, where dried larks' hearts sewn to silk ribbons and worn close to the skin to cure colic. The book instructs one how to use an eagle owl to trap other birds, how to preserve kingfisher skins to decorate and brighten your home, and how to feed young finches on homemade pellets pressed from walnuts and marzipan. It brims with delight and novelty: before reading it I did not imagine that one could train siskins to fly to one's hand at the sound of a rattle or bell.

The *Uccelliera* is also a part of a continuous tradition of European aviculture. Though Cassiano's world has gone, the bird species in his book have not, and the particular fascinations they have held for human keepers have proved robust. For example, Cassiano's passionate interest in songbirds with unusual markings reminds me of when I watched a group of men at a bird fair some years ago bidding hundreds of pounds for a goldfinch whose plumage was splashed with patches of pure white. It was a bird that Cassiano would also have

prized. His book shows us that seventeenth-century birdkeepers were fascinated by songbirds that could be taught to mimic the songs and calls of other species. That fascination is still alive in modern birdkeepers, who train their own canaries and hybrid finches by playing them YouTube videos and digital audio files of other birds.

Most soberly of all, the *Uccelliera* reminds us of the natural treasures we have lost, and are still losing, from our everyday lives. 'In the countryside about Rome' Olina writes, 'an infinite number [of quail] are taken every day'. The sheer abundance of bird life that Cassiano, Olina, Leonardi and their peers would have encountered is almost impossible to imagine. In less than my own lifetime, 420 million birds have vanished from Europe's woods and fields and hills. The *Uccelliera* is a testament not only to vanished ways of cataloguing the world, but also to vanished ecological plenitude. And perhaps most movingly of all, it is a memorial to the individual birds portrayed by Leonardi: small creatures that sang and hopped and scratched their beaks in the sun nearly four hundred years ago, and whose descendants still fly today.

---

above: **A hunter disguised as a bull approaching a group of partridges.** Detail of an engraving in Olina 1622, fol. 56v

opposite: **European robin.** Engraving in Olina 1622, fol. 15v

1 D. Freedman, *The Eye of the Lynx: Galileo, His Friends, and the Beginning of Modern Natural History*, 2002, pp. 54–5

2 P. Findlen, 'Cassiano dal Pozzo: A Roman Virtuoso in Search of Nature' in *Birds, Other Animals and Natural Curiosities,* Part B. IV–V of *The Paper Museum of Cassiano dal Pozzo: A Catalogue Raisonné,* I, 2017, pp. 18–46

Pasta *for*
NIGHTINGALES

Of the
GREAT TIT

OR *Tom Tit* OR *Great Titmouse*

HIS LITTLE BIRD is called *Parus major*, and otherwise *Fringillago* [or Ox-Eye Tit], vulgarly called in Rome the *Spernuzzola*, in LOMBARDY the *Parussola*, in Tuscany (with a name most apt to signify its call) called the *Cincinpotola* or Tom-Tit. In Piedmont they call it the BLACKCAP, as the greater part of its head is black. There are, in addition to the above-mentioned, THREE other species, which differ from one another, one by having a long tail, whence it is called *Parus caudatus*, and one known in Rome as *Potazzina*... being on its head almost altogether white, with the rest of the body, for the most part turquoise or deep TURKISH-BLUE...

These first three species are in almost every country, seen there in any season, also where people have their DWELLINGS, in the gardens... It uses not only its wings but also its nails, with these gripping fast onto *walls*, and *trees*. It is a spirited little bird, that when it has its young it defends them from other Birds with great *courage*...

They are taken, either with the SNARE or the Trap-Cage, as we call it, or with the spread net, or in a trap, putting about it LIMED TWIGS, and covering the cage with Greenery, because being a little bird loving towards its kind, when it hears a *companion* call out, at once it flies to them, and so it falls captive... its Call is very often annoying, resembling that SHRIEK *or* SCREECH that is made when opening a *rusty latch*.

Great tit
Watercolour and bodycolour
over black chalk, RCIN 927621

Of the HOOPOE

THIS ONE SHALL also enter into our present discourse for the reason of caprice; because it is in fact among the most beautiful of Birds that we have. It is called, after the Latin manner, *Upupa*, and in the vulgar tongue, HOOPOE. Its body is no greater than a common Thrush, and it has on its head a Tuft of feathers, which it constantly *raises* and *lowers*, UNFURLING and furling them as it pleases...

It is said, that every year they change their *feathers*, and that this is the reason that for a CERTAIN TIME they are not seen. It Flies slowly and in its flight it seems that it goes in *skips* and *jumps*...

It feeds upon WORMS, Ants and *caterpillars*, and upon GRAPES in their season, with which it sates itself in such wise, that sometimes it finds itself dazed and *half-drunk*. To remedy this, as some have written, it takes in its mouth a frond of the herb *Adiantum* [MAIDENHAIR FERN]. Others say that it puts this same herb in its nest, as an Amulet for salvation, and the safety of its young. Innumerable *falsehoods* have been written by the Arabs of this bird, such as saying, that by BATHING the TEMPLES ➤

Common hoopoe
Watercolour and bodycolour
over black chalk, RCIN 927682

➤ CONTINUED

with its blood, one sees in sleeping *marvellous* things; that the eye carried upon one's person cures from LEPROSY; that its skin attached on one's head takes away the pain from the same, and various other *incredible* things...'

Bubbola

Common hoopoe (*Bubbola*)
Engraving in Olina 1622,
fol. 35v

Of the
QUAIL

HE QUAIL, WHICH in Latin is called *Coturnix*, is a Bird smaller than the *Partridge* by a half, or a little less; in its size, colouring and appearance it is similar to the latter in many things. In the Spring it is to be found among the GREENERY of the MEADOWS, in the Summer among the same grown ripe, and in those cut down, among the Stubble. It is a Bird of passage, coming to us from the *Levant* towards the beginning of April, and returning thither near to the end of *Summer*, or in mid-*Autumn* at the latest; yet many, being too plump, find themselves unable to cross the Sea, and remain in these our parts, making for the marshes of the MAREMMA...

They dwell always upon the ground, where they even nest, as has been said of the PARTRIDGE; it is said that they lay from fifteen to sixteen eggs. They feed upon *grain*, and divers pasturage; but more willingly than any other thing upon *Millet*, eating also the seeds of *Hellebore*, which is thought by natural instinct to remedy the FALLING-SICKNESS, from which they suffer due to humidity of the Brain... they are fattened in Pens, or Coops made for the purpose, where they should be given Millet and Grain, and sometimes *Hempseed*; have their water changed often, and their drinking-trough cleaned. They are taken in great quantities throughout all the places where Grain and Forage are cultivated. In the Countryside about ROME *near* NETTUNO, upon their arrival an *infinite* number of them are taken each day. ➡

➤ CONTINUED

Their eggs have the self-same use as those of the *Partridge*…
unguent is made of their fat together with other ingredients. They
are of the most EXQUISITE flavour, and consuming them in
moderation is good for the *blood*, but on the contrary over-eating
them when they are FATTENED to the utmost often makes the
blood turn *rotten*.

Due to their being disposed to GROW FAT, it is thought that
they do not exceed four or five years of life.

Common quail
Watercolour and bodycolour
over black chalk, RCIN 927669

Of the
FRANCOLIN
COMMONLY KNOWN AS *Franguellina*

ECAUSE THIS BIRD with the loveliness of its feathers is the *ornament* of the aviary, it will not be unbecoming to include it among these others... also because it is a RARE BIRD in these parts, so unless something is said of it here many people will have no knowledge of it. The above-mentioned Bird, which is of the species of the *Starna* (that is the Grey or ordinary PARTRIDGE, as we may call it), is called in Latin *Attagen* and otherwise *Perdix Asclepica*, and in the vulgar Italian tongue FRANCOLINO. It is thought that this is in allusion to the freedom and openness of its life, as it has been compared to outcasts, and those under banishment as outlaws...

And as to its traits and features, it is proportioned very like the Partridge, though a little BIGGER and of colours different from the latter, being on the breast and on the stomach all *speckled* with white and black; having the tips of the wings and the tail similarly black; the head, neck and rump are *lion-tawny* shading towards pale reddish, with some little IRIDES-CENT shimmering purple-violet and black; the beak and the legs black, after the exact garb of the Partridge.

They are found abundantly in *Barbary*, most of all near to TUNIS, whence some have given it the name '*Partridge of Barbary*'; similarly in Rhodes it is said that they are there in quantity, as also in Spain in the countryside where there is rosemary, and *Spikenard* or *Lavender*. In the Aviary of His Most Illustrious Lordship Cardinal Borghese, these said Francolins are to be seen. They do not sing, but they have a certain SCREECH or SHRIEK that is so lusty and strong that it can be heard a great way away...

They are of exquisite flavour, being by many preferred to the PHEASANT, whence St Jerome, finding a great Hypocrite in one place, mocked the latter's *pretensions* and his wish to make an impression by saying:

Black frankolin
Watercolour and bodycolour
over black chalk, RCIN 927684

Tu Attagenem eructas, & de comesto Ansere gloriaris, that is You BURP a Francolin and boast that you have eaten GOOSE. They are also good for the health and their meat is very well suited to those who have a weak stomach or who suffer from Kidney Stones...

If you wish to keep them in the AVIARY you must have there a small case or crate in which they may hide away and nestle down, putting in there also some piles of *tufa* or spongy rocks with some sand. Their food is mixed *bird-food*; the male differs from the female in being more charged in its colours. They live for as long as Partridges.

Of the
ROBIN REDBREAST

HIS BIRD IS almost as well-known as may be, with its simple name, declaring itself by that which appears most markedly in it, which is the RED BREAST... To keep it hale, it is well done to give it sometimes some *Grubs*, LITTLE WORMS, that are found beneath the dung, or the earth, or sometimes in Summer, flowers of *Spanish Broom* or Weaver's Broom, or *Gooseberry*, which will make it merry.

Because this Bird is very FINE *and* DELICATE, it is inimical to an excess either of heat or of cold, therefore in Summer it retires to the scrub and *brush* or to the Mountain where there is GREENERY, and cool freshness; and in Winter it comes close to the homestead and shows itself upon the *thickets* and in the KITCHEN GARDENS, most of all where the Sun strikes, which it diligently seeks out, stopping on the *highest spots* that are most exposed to the same. It brooks no COMPANION, striving with every exertion to chase away any that would disturb its possession, from whence is born the proverb, *Unicum Arbustum non alit duos Erithacos* ('A Single Bush does not feed two robins'). It is a friend of the BLACKBIRD, in whose company it is most often found, and on the contrary it is a *great enemy* of the Little Owl... It suffers from DIZZINESS, or EPILEPSY. It lives from four to five years, and sometimes more, according to the diligence with which it is kept.

European robin
Watercolour and bodycolour
over black chalk; red chalk,
RCIN 927626

Of our native
TURTLE DOVE

THE TURTLE DOVE is of three kinds. The first is our own native one, which is called simply *'Turtle dove'*, and in Latin TURTUR, a term invented in imitation of its song. The second is the white. The third is the *Indian*, that others call the dove of ALGERIA or TURKISH dove. The size of the first kind (which is common, and known to all) is little different from the *Wood Pigeon*; it is as fine and gentle as any *Dove*, and it is everywhere of a grey colour, but here light and here dark, with some admixture of colour such as of RUST or BAY. The second kind is all white, and more slender in the waist than the first. The third, that is the INDIAN kind — the female is all white save only the beak, which is *blackish*, and the feet which are *red*, but the male has the head, neck, breast, and the primary feathers of the wings of a colour between pale YELLOWISH and BAY. Turtle doves are in almost every country. They live matched *two* by *two*, and it is said that, when one is missing, the other takes no new companion.

The white Turtle dove is found in *Poland* and in cold, snowy places. The Turkish, or Indian, is brought often-times from *Alexandria* in EGYPT, and one and the other are domesticated in such wise that they breed at home, lay-ing two eggs per clutch every *month*, and this for up to four years, after which it is said they greatly DECLINE, and die... Being captured, they are put into the pen or coop to fatten with *Millet* and *Panicum* grass; their consumption is most beneficial to the scours and the FLUX, being most efficacious in the same. Its blood, reduced to a powder, has even better properties.

European turtle dove
Watercolour and bodycolour
over black chalk, RCIN 927675

Of the
LAPWING
and the Hunting thereof

HIS BIRD OBLIGES me to come away from the thread of our narrative, which is more about the Birds that *sing* than otherwise; but knowing what pleasure details of this nature may bring to readers and to the greatest professors of the HUNT, I shall dispense, as I have said, with ORDER; though we shall return to it nonetheless as soon as the matters of this *Chapter* have been dealt with.

The LAPWING, then, which in Latin is called *Capella* or VANELLUS, is a native bird of ours, with the body and gait of the Rock Dove or *Turtle dove*, but longer. It has a head and back of iridescent shimmering green and black, with a little TUFT of four or six little feathers that grow out of this *sloping part*, with two alone of these set a little farther off than the others. The body with the inner part of the wings is *white*; its NECK is furnished with feathers of deepest *black*, that make a sort of collar... those on the outside of the wings are of the very self-same colour as the head; the RUMP is bay, as is also beneath the tail, which is part white and part black, but in this the last two FEATHERS are absolutely white.

It is usually found on the *plains*, and marshy places where there is enough HEATHER, and near to lakes and streams, around which it stays more than otherwise, for the abundance of *Worms*, FLIES, *Caterpillars*, SNAILS and the like that are found there, on which it lives. In Summer oftentimes it is found alone, but in Winter the birds gather together, flying in *flocks*. Its flight is swift, and accompanied by a continuous noisy chirruping.

This bird is more ESTEEMED for its loveliness than for anything other; whence it is *customarily* kept in gardens where it is marvellously useful for DIGGING out of them the progeny of Worms and *Caterpillars*. ➜

Northern lapwing
Watercolour and bodycolour
over black chalk, RCIN 927686

➪ CONTINUED

They are also suitable for EATING, being of quite good flavour and *nourishment*, but because of their quantity they are not prized.

It is proper to hunt them from ALL SAINTS' DAY until St Catherine's Day, using 15 or 20 model birds and two living ones that serve as a lure, and which shall be given HEART to eat, cut into long strips after the manner of WORMS. On this occasion you must make the *whistling* call of the Bird, which may readily be COUNTER-FEITED with a pipe made from a little strip of *Vine*, folded in such a way that it is doubled over, and putting in a piece of VINE BARK for a 'tongue' or reed.

This Bird is similar to the PLOVER, which in Latin is called *Pluvialis*, which is taken in the same way, Lapwings being very often confounded with *Plovers*.

Eurasian golden plover
Watercolour and bodycolour
over black chalk, RCIN 927687

Of the
SISKIN
or *Barley-bird*

 HIS BIRD IS called in Latin *Ligurinus*, or more commonly *Lucarino* (that is, SISKIN), and by some with the Sicilian name of *Lecora* or Barley-bird. It is a Little Bird of the colour of the Serin, but somewhat more GREEN, with a patch of BLACK on the head, as in the illustration shown here. Its body is a very little bigger than the latter, with the TAIL shorter, and its back and wings *splashed* with dark, also like the *Serin*, but somewhat more sombrely. The Male is distinguished from the Female by the PATCH on the head, which is much more black, being also on the stomach, the breast and the rump more COLOURFUL than she. The young differ from the old by this same rule of the *brightness* of the colours, and thus similarly those freshly taken from those long *caged*; it happens that the younger they are, the more BEAUTIFUL and light are their colours.

No particulars are known of how they make their *nests* or of their laying and *brood*, as they do not do it in these lands, whither they come (some say) from GREECE, some from Hungary, and others from the lands of the Swiss, which is more probable, going by the evidence of the Writers of *Natural History* of those lands, who say that they are found in great quantities there, and most of all in the Summer, and that they make their NESTS in the woods and among the GREENERY. They arrive every three or four ➤

⇥ CONTINUED

---

years, coming very often in such quantities that some believe they are borne on the *Wind*.

Their song is delightfully varied, and for this they are ESTEEMED, but much more when they have learned the call of some other Bird, in which they readily succeed, counterfeiting the *Goldfinch*, among others, excellently well. In *Rome* many tame them and accustom them, because they are so pleasant and likeable, to be outside the cage and to come to the hand like a SPARROW-HAWK, which they do by keeping them hungry at first, then by showing them a NUT broken in pieces, which they have them eat on the hand, holding in the same hand a little *bell* or *rattle* so that with this they are tamed and ACCUSTOMED to return to the hand at any time they like.

They are taken in the Autumn with the *Clap-net* in their passing from the MOUNTAIN to the PLAIN, and the quantity taken is so great that it discourages them. They fly in *flocks*, and for the most part if one flies down, many are taken, because all do so. Those few that out-run the nets make for the woods and the *Maremma* at the beginning of Winter to flee the cold. In the countryside they live in the same manner as the GOLDFINCH, eating particularly the seeds of *Thistles*, being so continuously among the brambles that this has given them the name of *Spinus*; but in the cage ordinarily they eat *Panicum* or HEMPSEED. They live from eight to ten years.

Eurasian siskin
Watercolour and bodycolour
over black chalk, RCIN 927604

Of the
DANCING
BIRD
OR RATHER *Wagtail*

 HIS BIRD IS called in Latin *Motacilla*, from the continuous movement it makes with its tail: in Italy it is called by divers names: in Rome CODINZINZOLA (that is, *Zanytail*), or BOVARINA (that is, *Ox-follower*, for its habit of following the oxen at ploughing); in Tuscany Cutrettola (that is, WAGTAIL or *Trembletail*), and in Lombardy BALLARINA (that is, *Dancing-bird*). Its traits and features are, as to the body, of the size of a *Warbler*, with the tail two times as long; the beak extremely fine, and black. There are two sorts of them found. The first is white and black, and is called *Motacilla alba*, or the White Wagtail. The second is yellow-green, MOTACILLA FLAVA, or the *Yellow Wagtail*... ✒→

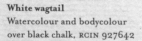

White wagtail
Watercolour and bodycolour
over black chalk, RCIN 927642

Yellow wagtail
Watercolour and bodycolour
over black chalk, RCIN 927647

➤ CONTINUED

They are found near WATERWAYS and the banks of rivers
and ditches, tracking *flies* and *maggots*, in which respect they are
also very often to be seen where ploughing is done, or where
there are livestock, whence it has acquired the name of BOVA-
RINA. *Bellone* in his observations calls it the *Culicilege* or 'mos-
quito-catcher', which if it were true this Bird would be praised
as the EXTERMINATOR of the most importunate creature there
is among all the insects... It does not survive for long in a cage,
nor does it sing there, although in the countryside its call is not
ill-favoured, most of all when it is *cocky* and crows boastfully
for having escaped the tempestuous attack of THE HOBBY. The

Grey wagtail
Watercolour and bodycolour
over black chalk, RCIN 927648

white kind is not seen here among us save in the *Autumn* and the *Winter*. The other is also seen in the SUMMER. The latter nests in cool places, sometimes making its nest upon the roofs of country houses: one and the other in Winter risk coming near to our HABITATIONS, letting themselves be seen in the gardens of houses and in our courtyards.

It is written, being *dried* in the oven, as has been said of the *Lark*, with equal parts of DEER'S BLOOD dried in just the same way, that taking one *dram's* weight of this composition with saxi-frage-water or strong WHITE WINE, on an empty stomach for a few mornings will have the power to break or reduce *kidney-stones*. Those wishing to rear them should tend them according to the rules for tending the NIGHTINGALE.

Observation
concerning the
OX-FOLLOWER
or
WAGTAIL

N THE COUNTRYSIDE about *Rome*, where the nets are spread for this Bird continuously from mid-*August* throughout all *October*, they maintain that there are some differences between this and the bird we call CUTRETTOLA (that is, *Tremble-tail*) or BALLARINA (that is, *Dancing-bird*); this one being properly Rural and following the Oxen, whence it has acquired the name *Bovarina* or Ox-follower. These are seen in two colours. One is clad in grey mixed with dull green with some little white in the primary feathers and in the Tail. The other grows quite yellow like the MOTACILLA FLAVA (that is, *Yellow Wagtail*). Both one and the other have very fine slender legs, which are black, as is also the beak. They feed, in addition to the FLIES that they take around the Cattle, upon the fruit of *Brambles* and on MAGGOTS.

These birds are caught with the *Clap-net*, and from the first, for want of DECOYS, the *Little Owl* is used, and any Little Bird that flits about. Later this is done with two or three or more decoys of the same sort, contriving however also to use as well as this the WHISTLE, with which their call is *counterfeited*.

The nets are spread early in the morning, and late in the evening near to some *Canebrakes* or Groves of REEDS, or where there are CATTLE or large Livestock; with these some other little

Grey wagtail
Watercolour and bodycolour
over black chalk, RCIN 927616

birds are also taken, and particularly one that is quite like the *Lark*,
most of all in the Legs and COLOURS although in the look of the
waist it is more slender. In the same countryside around Rome it
is called SPIONCELLO (that is, *Water Pipit*).

The *Bovarine* or Ox-followers are birds of passage: they are
seen in August, September and October, and are followed by the
BALLARINE or *Dancing-birds*, which from the end of October
tarry on the plain until the middle of Spring.

Of the
LINNET

WE MUST TAKE up this discourse once more to speak generally of the nature and kind of this Bird, which is commonly known as the LINNET, and in Latin *Linaria*, because it feeds upon Linseed; others would have it that it should be called *Salus*.

It is a little lesser in size than a SPARROW, but much finer in its build, and more slender, having a little round head, with the beak quite short and in proportion. It is all *earthen* colour throughout, being a little LIGHTER on the breast than on the back, and all *peppered* with droplets of the same colour... Our own COMMON kind is a little *bigger*, and the male is for the most part marked upon the breast with a few little patches of BRIGHT red colour, and this same patch is seen on its *head* above the stem of the beak. The *female* is without this, having instead of it others of a DARK UMBER earth colour. Our own ones have in the self-same way the tips of the wings and tail, as black divided with white. Having a red breast is not a *characteristic* of the different kinds, but rather a sign of age; it happens in the males that the OLDER they are, the more they are seen with the said parts covered with this same colour.

They sing very *sweetly*, both by their ordinary natural call and by being taught. They are REARED as has been said elsewhere, and they suffer from consumption, that is the subtle or WASTING illness... They are usually ill during the Dog Days, when many lose their song, and part of their feathers; this happens close to the *middle of July*. ✒

Common linnet
Watercolour and bodycolour
over black chalk, RCIN 927608

➸ CONTINUED

It nests in trees, not very *high*, laying three or four eggs. It eats the same things as the GOLDFINCH, its feed being ordinarily Canary grass seed, *Linseed*, *Hempseed* and Panicum grass. In HOLLAND they give them, as well as these, CABBAGE seed and *Rapeseed*. They are taken with the clap-net, most of all in the Autumn with other Birds passing through. It lives from five to six years.

Common linnet (female)
Watercolour and bodycolour
over black chalk, RCIN 927618

Of the
BLACKCAP

HE BLACKCAP, IN Latin called *Atricapilla*, is amongst all the little caged Birds by nature cheerful, with a pleasing, DELIGHTFUL song, and being lovely to look at, as it is very prettily divided into *light* and *dark* across the whole of its body. The head, the back and the primary feathers of the WINGS plus the tail are of blackish colour, but with a patch on the head absolutely black; the *upper side* of the wings tends towards green as if mixed with EARTH COLOUR; the body goes a little towards *yellow*, the underside tends towards white; and it has a *black* beak, a little HOOKED at the tip. It nests twice a year: the first in the last part of May, and the second in *August*, in saplings and in ivy hedges, *bay* or *laurel*; sometimes it nests earlier, sometimes later; it makes its nest of very fine roots and grass and also of the down or threads of the CLEMATIS called traveller's joy or old-man's-beard, or vines — whatever is conveniently close by.

It has *three*, *four* and *five* chicks. It flits and skims readily through the scrub, continuously uttering its call in the Spring. The good ones are the JUVENILES caught with Spider's-web nets; as soon as they are caught the tips of their wings shall be *bound*; and they shall be reared on heart in the self-same way as has been said elsewhere. They have a *woodland* call, and they will catch on to other sorts of calls learned from the LINNET, or indeed from other Birds, learning as nestlings everything that can be taught to them. This Bird requires particular care in being kept clean, otherwise it falls into MELANCHOLY and gets *sore feet*, dying of it if it is not remedied in a few days.

It is a wonderful thing to see, how this little bird is endowed with a particular KNOWLEDGE of its owner more than any ➤

Eurasian blackcap
Watercolour and bodycolour,
RCIN 927610

➵ CONTINUED

other person, giving a particular sign of this by singing differently when its owner appears near its *Cage*, and by continuously fluttering its wings, flying down to the bottom of this Cage and coming as close to the bars as it may. Some give them CHESTNUT flour, or tie a chewed *dried fig* to the bars.

In CAPTURING this bird there is sometimes a fundamental mistake made between this and the *Warbler*; which it is important to understand. The difference between them is this: that the Blackcap has the inside of the mouth BRIGHT RED in colour, while the inside of the Warbler's is *yellow* in colour. Many are caught out in this. The bird lives from five to *six* years if it is kept well.

Capinera.

Eurasian blackcap (*Capinera*)
Engraving in Olina 1622,
fol. 8v

Of the
HEMP
FIG-PECKER
or
BABBLING WARBLER

 HERE IS LITTLE writ-
ten on this Bird and it
is not known to many,
being taken simply for
the *Fig-pecker* or GARDEN WAR-
BLER. In Lombardy there are more
of them than elsewhere because of the
Hemp that this Province produces in
great abundance, amongst which this
Little Bird stays almost continuously,
flitting about and singing; it is called by
some 'Reed-sparrow' or TITLING, or
Reed Warbler. It is placed in the class
of Fig-peckers, as has been said, both
for the resemblance it bears them in its
traits and features, and also for its FAT-
NESS, and for this reason it is called by
some in Latin *Ficedula Canabina* (that
is, BABBLING WARBLER).

In its appearance it is similar to
the Fig-pecker and to the *Nightingale*;
to the Fig-pecker in regards to its size,
and to the colour it has on the STOM-
ACH *and* CHINSTRAP, being of a dull
green tending towards yellow; on the
rump, neck, wings and head it is of a
colour similar to the Nightingale, and
also in the tail, which tends to patches
of *rust-colour*. It makes its nest in the
Hemp, interweaving it with the down
or threads of *Clematis* that is called
Traveller's Joy or OLD-MAN'S-
BEARD and the tendrils of the same;
nesting sometimes in *thickets*, and in
among saplings, or in some dense,
thick thorn-bush, laying four or five
eggs, but for the most part *four*.

If you wish to rear it from the
nest, it is necessary that it be *fledged*,
hand-feeding it with MINCED
HEART, as has been said before,
offering some to it with a stick for ➤

Unidentified warbler
Watercolour and bodycolour
over black chalk, RCIN 927634

Probably wood warbler
Watercolour and bodycolour
over black chalk, RCIN 927603

⇛ CONTINUED

a few days, until it begins to *peck* for itself. Its food is the same as that of the Nightingale, which it also resembles in the FLATNESS of the head, and the fineness of the beak.

In its song it has several calls, similar to those of the *Blackcap* and to the Nightingale, whistling very sweetly. The Male has more RED on the back than the Female. It has been observed that when it *moults* its feathers, if it does not have the comfort and convenience of bathing, it DIES; therefore it is best at this time to spray it lightly every day, or to keep a *vessel* in the cage for this purpose, putting it then to DRY *in the* SUN. It lives from eight to ten years.

Beccafico ordinario

Beccafico Canapino

'Figpeckers'
Engraving in Olina 1622,
fol. 10v

Possibly whitethroat
Watercolour and bodycolour
over black chalk, RCIN 927635

Possibly garden warbler
Watercolour and bodycolour
over black chalk, RCIN 927611

Atlantic canary
Watercolour and bodycolour
over black chalk, RCIN 927609

Of the
CANARY

THAT IS THE
Canary Island Sparrow

HE CANARY, OR Sparrow of the Canaries, in Latin *Avis Canaria* or *Passer Canarius*, was not known to the ANCIENTS, whence little or no account of them can be had; therefore we shall include here *everything* from our own experience and from the reading of the MODERNS who have written thereon.

This Bird has been carried to *divers* places of Europe with the Ships of the CANARY ISLANDS, otherwise called the *Fortunate Isles*, from the happy temper of the air that is enjoyed there. Their traits and features wholly resemble the *Serin* and the *Siskin*, being however somewhat bigger than either. It is also different from the Serin, as the Canary has its BREAST all of one colour, that is of dull *Green*, tending a little towards yellow, and the Serin having its breast (although verdant) yet with more *yellow*… on the breast it is lightly SPECKLED with dark patches, as if with droplets of dark grey or UMBER earth colour, as the *Painters* say, being in this same way splashed about the eyes, or on the cheeks.

The Male is PRIZED for its song, and is distinguished from the female in this: that it is more yellow about the *breast*, *chin* and the top of the head than the female, which tends more towards GREEN. Among males, the best are those that have *longer tails* and less body; indeed, it has been observed from long practice that the finer in build they are, the BETTER disposition they have to sing; whereas those that have a bigger body, and have the habit of turning about in the cage, TWISTING their head, very often ➤

➥ CONTINUED

turn out to be *Tree Sparrows* of the Island of Palma, and of *Cape Verde*, which are not valued for singing.

There are also *Canaries* here of our own, said to be true Canaries, a great QUANTITY of which, in being brought by a Ship from those parts to *Livorno*, were by chance shipwrecked near ELBA; the cage being *dashed* in pieces in that *disaster* the birds saved themselves and survived upon the island as being the nearest land; they took REFUGE AND SHELTER and bred there, multiplying in such a way that now they are seen also in other parts. They have, however, with the DIVERSITY of the country, changed a little in their traits and features, becoming a *bastard sort* with black feet, and quite a deal more YEL-LOW in the chin than the legitimate Canary; besides this, they are the size of the *Siskin*.

Their feed is *Panicum* grass, HEMPSEED, Canary grass or *Millet*, but of this last they should be given more sparingly. The true Canary grass, which was brought with the birds from the Canary Islands, is *Phalaris*; this is

their proper food, which also comes in good abundance from SICILY to GENOA. Some have ventured to give them *Poppy* seed with good results.

To entice them, and dispose them more to sing, some CODDLE and *pamper* them with crumbs of Sugar, or a mite of *Sugarcane*, very finely pounded; others cover the top the cage with CHICKWEED (in Latin *Auricula muris* or *Morsus Gallinae*) for greenery, this being a herb dear to all the Little Birds that sing.

They sometimes suffer from SWELLINGS on their head, in which case care must be taken, anointing such either with *Butter*, or unren-dered *Chicken lard* until it is RIPE, then opening it deftly with the tip of a pair of scissors and cleaning out the *rotten* stuff; this will get the sore to HEAL over; otherwise they suffer from *bird lice*, and this is to be reme-died by spraying them, when it is not very cold, gently with *strong wine*, and exposing them then to the SUN, or in a warm place. They live from ten to *fifteen years*, according to the care taken of them.

**Citril Finch (identified in
inscription as 'canary of Elba')**
Watercolour and bodycolour
over black chalk, RCIN 927602

Corn bunting
Watercolour and bodycolour
over black chalk, RCIN 927654

Of the CORN BUNTING

THIS BIRD IS better known in the *Countryside* of Rome than elsewhere – quantities of them are found there. Some say that it should be called ZIVOLO MONTANINO (that is, *Mountain Bunting*), although it is different from the BUNTING both in its size and its colouring. In Latin it is called *Emberizza*. It is the size of an ordinary Lark, from which it is also not far off in colour, being all above the colour of UMBER EARTH, and below tending towards white, speckled with dark. It has a short, *stout beak*, with the upper part being thick and solid from the inner side as far as the middle, and is tipped with a *bulge*, or rather SWELLING, like that in the illustration placed here; with this it breaks open grains of *wheat*, *oats*, or other FODDER... it has legs like the *Lark*, but without the length of the back claw. Its song is a strident shriek, whence it is called STRILLOZZO, which is what the vulgar people of Rome term a *shriek* or a *scolding*... Its ordinary call, CHEEPING or PEEPING, is like that which is heard in the fields from the *Grasshopper*.

It nests on the plains, on the ground, like the Lark, or at most in some *thicket*, laying from five to six eggs. It feeds in the Countryside upon DIVERS SEEDS *and* MAGGOTS, eating also very willingly of wheat *grain* and *barley*. It is usually found on the ground, enjoying the plain more than any other. It is caged by

➤➤ CONTINUED

Fowlers for use with the clap-net, with which in the Autumn other Birds are taken; when caged they are given MIXED BIRDFOOD, and are kept in a low Cage without partitions or *twigs* or *perches* to get up on, like the above-mentioned Lark, which it does not exceed in age.

Corn bunting (leucistic)
Watercolour and bodycolour
over black chalk, RCIN 927660

Of the
BUNTING

HIS, LIKE MANY other Birds, is given a name from its call, in which it seems to say *Zi, Zi*. From this comes the name ZIVOLO. In Latin it is called *Cirlus*.

It is the size of a *Sparrow*, or a little larger, with a short beak, quite big, and its head the colour of the SERIN — green with light and dark above, DULL YELLOW round the eye and on the wishbone; on the back and wings a *reddish* colour, like bay, with the same reddish colour between the wings like a NECKLACE; the tail is an *indifferent* green; the breast and stomach tend towards yellow with some ADMIXTURE of green, sometimes splashed with *droplets* of dark colour. These ordinary traits and features of this Bird sometimes vary. One type is called the YELLOWHAMMER, or *Straw Bunting*, for the colour of straw shows in it quite brightly; the other, which is described above, is called simply BUNTING. Aldrovando writes that in Bologna this Bird is called *Raparino*, which it may be in those parts, but in TUSCANY that name refers to a completely different Bird.

It lives for the most part on the ground, PECKING about and searching for *seeds*; wherefore very often when it is captured, it is found to have its beak *ingrained* with earth. The birds go about in bands, often in the company of the CHAFFINCH, whose song in some parts they imitate. Because of this they are *caged*, as their song is not unpleasant, and they may also serve as a lure for the Clap-net. It is quite SIMPLE-MINDED, and can therefore be caught more easily than the Chaffinch, either in the Clap-net ➡

⇾ CONTINUED

or with BIRDLIME twigs. It is easily *tamed*, although for the first
two or three months after it has been caged it makes no other than
its ordinary call; it then settles down to SINGING quite success-
fully. It lives on *barley*, *millet* and PANICUM. In the Autumn and
in early Winter, especially on rainy days, these birds can be seen
in great numbers, going round NEW-SOWN FIELDS or freshly
ploughed land, finding *grubs*. It suffers from falling-sickness. It
lives about six years.

Cirl bunting
Watercolour and bodycolour
over black chalk, RCIN 927631

Of the
ORTOLAN

HE GOODNESS AND flavour of the *Meat* of this bird have caused observation of its Song to be forgotten, but although this is NOT EXQUISITE it may nonetheless pass among those that sing, and in *Lombardy*, where it is found in quantities, owing to the abundance of minor CEREAL CROPS, it is caged by some to sing and by others to eat, being apt and suitable to *satisfy* quite well in both these two senses, of *hearing* and *taste*. Therefore details of it must not be unwelcome... in its build it is no bigger than our own LARK, even a little smaller; for the rest it has a great resemblance to the *Corn Bunting*.

In the beak, legs and feet it is somewhat REDDISH, tending towards a dull FLESH COLOUR; the head, neck and breast tend towards yellow, with a few *sprays* or *splashes* of the colour of saffron; its stomach the same colour, with some patches of green. The PRIMARY FEATHERS of the wings and of the tail tend towards black, the remainder being between *yellow*, and *black*. The female differs from the male in being in the head, neck and breast yellow mixed with green, both colours being GLIMPSED distinctly; the male also has below the eyes on the sides a bright patch of colour, like the YOLK of an egg, which is not there in the female; it has the farthest toe of the foot (I mean the one behind) quite large, which serves as an *indication* that this is a Bird that goes upon the ground.

It is found in the countryside, in divers parts of ITALY, particularly in *Tuscany* and about *Bologna*, where it forages ➥

Ortolan bunting
Early seventeenth-century artist.
Watercolour and bodycolour
over black chalk, RCIN 927622

➤ CONTINUED

on barley, millet, PANICUM and the like, making its nest in the same, with *five* or *six* eggs. With care, and using an enclosure, it fattens so well that some come to weigh THREE TO FOUR OUNCES, whence being exquisitely delicious they are sent preserved in their skins, and rolled in FLOUR, to be served up in Rome and elsewhere to noblemen.

Ortolan bunting (*Ortolano*)
Engraving in Olina 1622,
fol. 21v

Ortolano.

Instructions for

AN ENCLOSURE
FOR *Ortolans*

OU MUST PRIN-
CIPALLY have a
care that in the
room intended for
this purpose, in a place not too
exposed to *Winds*, there should
be only enough LIGHT so that
the birds can see where to eat
and drink and perch; secondly
that where the light comes in
they should not be able to see
GREENERY or *countryside*, so
that the desire for these should
not make them *melancholy*; third,
that in their drinking-trough
the water and the vessel must be
CLEAN, perhaps with a *fountain*
fed via a little channel; fourth,
let the gateway of the ENCLO-
SURE be small; and fifth, let
the room be extremely well
plastered or daubed, to make all
safe from *Moles* and other animals (and as well as plastering

it, it should be painted grey in
colour); sixth, in every cor-
ner there should be a pole full
of little branches to serve as a
PERCH, a little distant from the
wall and with the perches get-
ting smaller after the manner of
a *rack of shelves* in a dresser.

Have another little room at
the side of the enclosure, which
opens when you wish to take some
of the birds out, allowing the
number needed to enter there
without DISTURBANCE, and
which can be closed off by pulling on a length of *twine*, so that
the others do not take offence
and grow *melancholy* upon seeing their companions taken and
SLAUGHTERED...

These Little Birds live from
three to four years, often dying
before by reason of their *excessive*
or *immoderate* FATNESS.

Of the
BRAMBLING
or
MOUNTAIN CHAFFINCH

Brambling
Watercolour and bodycolour
over black chalk, RCIN 927644

HIS GOES UNDER the same name as other *Chaffinches*, differentiated by the addition of the place where it lives, which is the MOUNTAIN, whence it is called Mountain Chaffinch and in Latin *Montifringilla*. It is no bigger than a SPARROW, with a beak quite *large in size*, and *sharp*; its colour tending towards yellowish, and blackish at its tip; its head, neck and back change between *black* and *rust*-colour; on its rump it has a little bit of WHITE; the tail is black, with *two feathers* on the sides part white and part black; beneath the throat the CHINSTRAP is black, and on the breast it is between red and yellow; the *stomach* is white; the wings are black striped with two transverse bars of colour: one, that is the first, *reddish* and *yellow*, the other white; the legs are a little BIGGER than those of the ordinary Chaffinch.

The female is known by being more charged with this said RUST-COLOUR, with much less *black*; being also beneath the eyes and on the throat and breast less brightly coloured than the male.

It is a BIRD OF PASSAGE, and arrives in the *cold*, later than any other. It is seen *particularly* when it is very cold, and when there is snow. The FOWLERS cage them, more to use in the Clapnet than because they sing *exquisitely*, for they do not make more than an ordinary call, save for one call that seems as it were the mewing of a CAT; however, if it is kept close to other Birds, it steals something from them, in particular from the *Sparrow*, so that in a short time it imitates them perfectly, and thus sweetens its peculiar song and is made EASIER in itself by the presence of others, so that in no time at all it is no longer so sad ... they go about in bands, and feed upon divers *Seeds* and on MAGGOTS, as has been said of the others.

They are kept in the Aviary for their *beauty*. In the COUNTRYSIDE about Rome they are quite rare. Their food in the cage is *panicum grass*, and *Hempseed*.

Of the
HAWFINCH

HE FRENCH NAME for this Bird is considerably better known than our *Italian*, because they call it GROS-BEC, which is to say *Grossobecco* or Gross-Beak; in Italy it is called Frisone and also *Frosone*. Others, in imitation of its French name, mean this one when they talk about the NUTCRACKER, because with the hardness of its beak it breaks open the pits of *Cherries* and *Olives*, and of these in great part sustains itself. In Latin it is called *Coccothraustes*. It greatly resembles the CHAFFINCH in its plumage, most of all in the colour of the wings, but it is bigger by a third, and shorter and fatter in shape. Its head is somewhat bigger than theirs, proportioned like the body; its beak is LARGE and SHORT, and at its base so broad that it forms almost a triangle. Round the *eye* and below the Beak it is contoured with BLACK, while on its head it is *yellowish* in colour, tending towards red; the slope of the neck and about the neck on both sides is *greenish feathers*; the back is of dark bay; the end of the tail grows in white. It is found in the summer amongst the woods, or in the MOUNTAIN, coming down in October to the plain.

It nests in the HOLES OF TREES, laying there five or six eggs, and feeds on *divers* seeds, particularly of *Hemp*, eating also CHERRIES, OLIVES and diverse *berries*; breaking open the pits and eating the kernels thereof. It also spoils the eyes of Plants, as has been said of the *Finch*... It should be tended as has been said with seeds of *Hemp*, Panicum grass, CANARY GRASS and the like, and they are good to eat but do not do well in *Aviaries* if they are very small, because in that case they annoy the other Birds. It is not held in esteem as a SINGER. It lives as long as a Chaffinch — a little more, a little less.

Hawfinch
Watercolour and bodycolour
over black chalk, RCIN 927657

Of the
FINCH

THIS BIRD IS called by some *Finch*, and by others *Chaffinch*. Those of Bologna call it SUFLOTTO, that is the *Whistler*, so named it is thought because in its song it imitates a REEDPIPE or end-blown *Flute*; it is also called by some Fringuel di Montagna or *Fringuel Montano* (that is, Mountain Chaffinch or BRAMBLING); some, for its varied piebald markings above and below, have called it Monachino or *Little Monk*; however this may be, it is well known in almost every language by many different names: one thing certain is that in Latin it is called RUBICILLA (that is, *Rosefinch*), from the redness of its breast, or *Pyrrhula* from the Greek (that is, *Bullfinch*).

The Finch is a very BEAUTIFUL Bird to look on, being a little bigger in its build than an ordinary Chaffinch; it has a short, wide beak, a little *hooked*, black and *lustrous* like Pitch; its TONGUE is quite long, and wide; its back is sombre in colour, tending towards *Turquoise* or TURKISH-BLUE, being black upon the Head, Tail and the tips of the primary feathers of the *wings*... The throat, the breast and the stomach are the bright colour of *Minium* or Red Lead, or the POMEGRANATE flower; beneath the wings it is white; its legs and feet are *fine* and *black*. The male differs from the female in being red, as has been said, while the other, in the same parts, is the colour of CHESTNUT, blended with grey.

Writers very often fall into an *error* and mistake it for the *Robin Redbreast*, because that one is called RUBECULA and this Rubicilla, but the differences are very clear to see, the red of ➥

Eurasian bullfinch
Watercolour and bodycolour
over black chalk, RCIN 927670

➺ CONTINUED

---

the former tending towards the COLOUR OF RUST and of this latter towards *Minium*. The latter is of the build of a Warbler or *Fig-pecker*; this one is bigger, with a large head and a long and hooked beak; that one is seen in the WINTER and this in *Summer*, so there is little excuse for confusing them.

It dwells always in the Mountains, and is found particularly in those of BOLOGNA and of MODENA, but sometimes in Winter it comes down to the *Plains*. It nests among thickets, laying four eggs. In the Countryside they feed upon GRUBS, *Hempseed*, and some *berries*, and in Spring they cause no little harm to divers fruit trees, most of all apple- and pear-trees, most willingly eating the heart out of the HARVEST they make... They are simple to put at ease, so that in the *Aviary* and in houses they nest and *brood*. They learn to COUNTERFEIT whatsoever call or cry may be desired to be taught them; also the calls of any other Bird; and some have even learned a few words. The female SINGS NO LESS than the male, which is *unusual*.

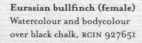

Eurasian bullfinch (female)
Watercolour and bodycolour
over black chalk, RCIN 927651

European goldfinch
Watercolour and bodycolour
over black chalk, RCIN 927619

Of the

GOLDFINCH

 MONG THE BIRDS that sing, the *Gold-finch* has a signal place. In Tuscany it is called CALDERUGIO (that is, *Thistle-finch*), and in Latin *Carduelis*, and were it not for fact that they are so abundant, they would be held in GREATER ESTEEM than they are, being not only beautiful, but also *songsters*. The Male is the best as a songbird, being able to learn very readily calls from the *Linnet*, the *Blackcap* or the CANARY, whence it then forms a mingled song that is very sweet to hear. This however can be done only with NESTLINGS, or with those newly fledged. The male is known by several signs: it has a *longer, bigger beak*, its head is splashed with black and bright red, the primary FEATHERS of the wings being tinted similarly as far as the mid-point with deepest *black*, and with the most lively bright yellow. The wings of the female are quite *greenish*... in the nestlings or FLEDGLINGS these colours cannot be recognised.

To rear them, they must be had from the NEST at the time when they are already well *fledged*; and they should be fed in the following manner. You must take *ciambellete*, or little CAKES, peeled *Almonds*, and *melon seeds*, pound them all together, and make a paste of them: the same may be done with WALNUTS and a little *Marchpane*; of this mixture, little pellets are made, like the seeds of *Vetch*, which you will offer to them with a stick, one at a time, giving them three or four per CHICK; this being done, you will have at the other end of the same stick a little COTTON-WOOL, this being *plunged* in water it will be offered to each one, dipping it again each time... In choosing them be advised that the best are those of *August*, and those that are found in NESTS made among *Blackthorn*, and thickets, or upon ORANGE-TREES: they also hatch in May and in June. They suffer from *Vertigo* and also *Epilepsy*, CONSUMPTION and melancholy... They live from ten to fifteen years, according to how sound they are, and to the good *care* that is taken of them.

Of the
GREENFINCH

JUST AS WE have said of the *Robin Redbreast*, which by its name gives us to understand its TRAITS and FEATURES, so it can be said of the present Bird, which is called *Greenfinch* for being all green: it has its Latin name of *Chloris*, originating from the Greek word, signifying GREEN ... it is a Bird a little bigger than a *Sparrow*, all green in the part beneath, tending towards yellow, and above towards dark green, ADMIXED, as has been said before, with the colour of *Umber* earth that is called *Ochre*; it has a rounded beak, sharp, short and stout, and somewhat WHITISH at the end near the body.

The Female is much less colourful than the Male, showing little green. It nests in VALLEYS, and low places, making its nest sometimes in *Sallows*, among grasses, and most of all comfrey, stowing it with a CARGO OF WOOL and *hair*; it lays from three to four eggs. It sings quite sweetly, most of all in the company of other Birds. It is easily TAMED, becoming accustomed to come to the hand, and to cast out the *lees* of its *food*, and drink, most deftly indeed... Those who delight in using the CLAP-NET generally have some of these, because with their cry a great many are taken.

It feeds in the countryside on the seeds of THISTLE, Burdock, *Turnip* and Phalaris: in the cage it is given *Panicum* grass, Hempseed, and CANARY-GRASS. It lives from five to six years.

European greenfinch
Watercolour and bodycolour
over traces of black chalk,
RCIN 927627

Of the
CHAFFINCH

UR OWN NATIVE *Chaffinch*, which is so called in opposition to that of the *Mountain*, in Latin is called FRINGILLA. It is a *Little Bird*, the size of a Sparrow or a little bigger; it has a big strong beak the colour of FLESH, tending towards *black* at the end; the head and neck are of a colour that tends towards TURKISH-BLUE (that we call *Turquoise*): the back is of the colour of *Chestnut*, the rump greenish, the breast is between red and bay; both the wings are PIED, with two sides of white, being in the *middle* and at the tip black; the tail is of the same, save the TWO FEATHERS on the sides which are white.

The female is distinguished by being a little NARROWER in the head, and paler coloured, most of all by the *breast* being rather grey in more places than otherwise. One that is smaller and more slightly made or SLENDER will be more of a *songster* than others.

This one is counted as a BIRD OF PASSAGE, which is true enough, even if in spite of this some are always found in our parts; these however in the *Summer* make for the *Mountain*, singing some in a simple fashion with quite a SHORT CALL, and others with a long repeated call; these ones are prized for the *Clap-net*, so that they may serve as a lure. It nests in the Summer in the Mountains, sometimes in the OAKS, and sometimes among *dry twigs*, forming its nest of TREE MOSS on the outside, and of that *downy wool* that is wont to fall from some trees and plants. They have from four to five *Chicks* per brood.

The nestlings are reared by the same rule as for the GOLDFINCH, and other tiny *Little Birds*, and these, that is the aforesaid juveniles, are kept under the tutelage of an old, good *Chaffinch*, so that like this they will learn calls that are long and beautiful. As well as singing, they are TRAINED easily to draw their food to them, and to drink from a *little cup*, aiding themselves not only with the beak but also with their FEET. When it is desired that they should sing a lot, they should be given a little BREAD and chewed *sheeps-milk cheese* — that is to say cooked ➻

Common chaffinch
Watercolour and bodycolour
over black chalk, RCIN 927646

➥ CONTINUED

cheese, taking care that it should not be salted. Others give them, to this same effect, the WORMS that are given to the *Nightingale*, or some GRASSHOP-PERS. They arrive in these parts in the Autumn, when they are taken with the Clap-net in great *quantities*, and in the colder season some of them also with the LITTLE OWL.

It is quite a shrewd and *wily* Bird, so that if it should catch any glimpse or hint of the trap, whether Nets or *Lime-twigs*, no matter what call or lure may come from its COMPAN-IONS, it will not fly down again. The *Fowlers*, so as to have good songsters at that time, will keep them ENCLOSED in the Spring and in the *Summer* so that they do not pour forth their song. The Fowlers do not catch many of them because if they sense the WIND or anything that annoys them, they *do not sing*; and because they are, as was said, a SHREWD and WILY Bird, and *suspicious*, if there is not a continuous lure or ENTICEMENT they do not fly down. Its ordinary food is *Panicum*, and a few BLADES of GRASS; they are subject to growing blind. They live from seven to eight years.

A hunter disguised as a bull approaching a group of partridges. An enclosure can be seen in the distance (*Modo da pigliar le starne*). Engraving in Olina 1622, fol. 56v

Of the
NIGHTINGALE

IN THE PRESENT little work, if any place other than the first were given to this *Bird*, it would be openly to WRONG that which divers Writers of stories both *ancient* and *modern* have all agreed upon. Let them therefore have that place which the EXCELLENCE of their song and the *opinion* of many has given them. So to begin by expounding upon their TRAITS and FEATURES let us say that because of their russet colour they are called vulgarly in Tuscany *Rusignuolo* and *Usignuolo* (that is, Rosignol); in Latin LUSCINIA, the Etymology being taken from its singing in the *Woods*, that in Latin are called LUCI, or as others say from its song, which it makes before the first light of day.

It is a Bird of *passage*, and it is said that it comes each year from the LEVANT, arriving in our region towards the day of the *Holy Annunciation*, continuing to come until the end of APRIL, and then retiring thither at the beginning of NOVEMBER, or even earlier. Upon its arrival it takes a place for its own, as its own *franchise* or *freehold*, into which it does not admit other Nightingales save only its FEMALE PARTNER, and in this place it sings. It is found for the most part in cool shady places such as *Thickets*, tree-lined Alleys, *Hedgerows* and other similar places, where the trees are not very tall, delighting little in these save for the OAK. It broods in *May* or *August*, making its nest in a Bush in the scrub or Woods, and fitting it out with LEAVES, Straw, Clematis (traveller's joy or *old-man's-beard*) and Tree Moss, with four or five eggs. It won't sing near this, for fear of making it known, but for the most part sings a stone's-throw off.

The AUGUST BROOD is esteemed the best, for at this time this Bird has the *complexion* of its Humours hotter and drier, ➥

➤ CONTINUED

for which reason many have *preferred* or *favoured* those of the
Mountain over those of the Plains; the most preferred are those
from damp and MARSHY PLACES, since this may cause relaxation of those parts that are the instrument of the *voice*. They must
be had from the nest, well clad in *feathers*, when they should be
put in the bottom of a WICKER BASKET for wine bottles, made
of straw, with their nest itself, or with some straw or *hay covering*
it, so that they do not leave it. Fold their *legs*, and keeping them
ISOLATED at first, hand-feed them *eight* or *ten times* a day, on
Bullock-heart or raw Veal, well cleaned of SKIN, nerves, and fat,
making of it little pieces of the size of a *writing quill*, giving two or
three little pieces to each one, varying this sometimes with hard-
boiled EGG-YOLK. Give them drink two or three times a day with

Common nightingale
Watercolour and bodycolour,
RCIN 927607

a bit of *cotton-wool* on top of a stick, dipped in water. Continue thus and keep them COVERED until they begin to stand up well on their legs; at that time put them into a *Cage* with new straw at the bottom of it, still TENDING them as above, until you see that they wish to peck for themselves.

When they are reared, GRAINS OF PASTA, made as described below, shall be put in with them to one side, in *little boxes* or *drawers*, and on the other side heart, as above, spread out on a little square TABLET of stone, which shall also be put into a little BOX or *drawer* to hold it properly in place.

Among nestlings the males can be known by the fact that having eaten, they WITHDRAW to the top of the cage, and begin to *chirp* and *chatter*, moving the CHINSTRAP; the young female does little or nothing; also the male stands still sometimes for a good length of time UPON ONE FOOT, and at other times it will ➺

➺ CONTINUED

suddenly dash *furiously* several times about the Cage. Everyone agrees that these NESTLINGS do not sing as well as the *woodland* ones, as it is only fitting that the Father and the *Mother* should teach them how to sing, wherefore the WOODLAND ones are better than the others, and for this reason in order to make them sing well they should be *kept close* to one that has the true woodland call. Experience however convinces me that this observation is FALSE; that these succeed quite as well as those others, and it is the *gift of nature* that without TUITION they make the call that is proper to their kind.

Among the Woodland ones, the MALE is different again in that its eye is bigger, the head *rounder* and a little larger, the beak longer, the legs BIGGER, the tail longer, and a little or even considerably *brighter in colour*. In the Spring, it is easy to know them from the swelling (because they are going about their AMOURS) that is seen in their *sex*. To tell them apart from the *Redstart*, which when fresh out of the nest can be difficult to know from the Nightingale, you observe their call, that of the NIGHTINGALE being of such wise that it seems to says *ziscra* or *ciscra*.

The way of TENDING and REARING the Woodland ones will be explained in the following chapters; suffice it for now to say that if they are shy and backwards in *refusing to eat*, put grubs or MAGGOTS such as are found in Bran, or Flies, into a cage that is closed off, without *perches*, tying sometimes some of these grubs to little pieces of Heart to accustom them to eat meat.

It is said that the ANTS' EGGS serve them for *medicine*. In Germany and most of all in Nuremberg, the Countryfolk bring so many of them to sell that they are measured in *Quarter-litres*, as is done with PANICUM grass in our parts.

This Bird sings best all of *April* until *mid-May*. In Summer there are few that sing, either because of the MOULTING of their feathers, or because they suffer from the heat. Of the *nestlings*, most sing in the Autumn, and sometimes in Winter, if they are kept in a warm Room, or some place where the air is TEMPERATE. They suffer from *excessive fat*, in which case you must endeavour to slim them down, giving them two or three times a week some GRUBS either born out of Bran or beneath *Dung*, but not exceeding more than two or three at a time. If on the contrary they should become too thin, give them FRESH FIGS if in season and not dried, *plump* and well chewed. They grow so at ease that not only do they become accustomed in the cage to come with *charm* and *graces* to the finger, but even to remain outside of it, in which case they eat of any stuff, keeping them only from SALTED THINGS.

Common nightingale (*Rusignuolo*).
Engraving in Olina 1622, facing fol. 1

TO MAKE
Pasta to feed
THE NIGHTINGALE

FIRST YOU MUST take sieved *Chickpea Flour*, two or three pounds; half a pound of *almonds*; four ounces of butter; and four EGG-YOLKS boiled, and mashed; and after this shell the almonds and mash finely, you will take the above-mentioned things and mix them together, and knead them together with the flour of the said Chickpeas in a Basin after the manner with which SUGARED ALMONDS are made...

Taking a pound of *honey*, and three ounces of Butter, putting them in a new cooking-pot to melt, and having thus MELTED, and well boiled, the one who is in charge of the pasta having a *flat spoon* in hand, and another having a ladle with one or TWO HOLES pierced in it, thus taking up the honey little by little and *pouring it* onto the pasta it will come out better through the said holes: and that other person will KNEAD constantly, until such time as the said paste seems to you that it is [all] mixed in, and become *granular*, and this serves for the Summer. In Winter you must also add a COPPER BAIOCCO coin's worth of saffron, to be warm, & appetising, and you will keep the Bird happier...

When the above-mentioned pasta is become GRANULAR and yellow in colour it shall be taken off the fire; & having a *Sieve* made of round holes, you will force it with your hand to pass through... putting it then upon a table covered over with a white TABLECLOTH, you will spread it out the sooner to make it dry, and when it is dried it is put in a jar, or a *box*, and so you can use it to feed the said Nightingale.

opposite: **Weighing, sieving and cooking pasta to feed to nightingales.**
Engraving in Olina 1622, fol. 4v

TO ENCOURAGE
the Nightingale
TO SONG

IF MEN HAD that exact knowledge of *Nature* that might be obtained by long OBSERVATION and CONTEMPLATION, so much could be known that, with the right and proper means, it might sometimes be given to us to *understand* how Nature can be forced, seeing our *pleasure* at any occurrence out of the ordinary. Whence comes about the MARVEL of how Plants increase and multiply, of the changing of *colours* in flowers, of having these same flowers at any season, of having them with STRONGER scents, and fruits with qualities either healthful or NOXIOUS, and a thousand other bizarre and *curious* things, not only in these but in infinite other things, as is seen in the writings of *Francesco Giorgi*, Roger Bacon, Giovanni Battista Della Porta and DIVERS GERMANS. Whence it shall seem no miracle that with art one may bring the Nightingale to sing, either more than usual, or out of *season*. Thus in Winter giving it along with its Pasta some ground *Pine-nuts*, and in its Drinking-trough a thread or two of SAFFRON, as these two things both heat and cheer it, so without any *harmful effect* this will induce it to sing. Also infinitely effective is the *Sympathy* that this little Bird has with SYMPHONY and music. Therefore in the Room where it is kept, if you make a sweet Concert of *sound* or of *voices*, it will be wonderfully fired up to sing.

We see the like happens also in PARAKEETS, that being used to *chatter* and *talk*, if they ⇢

opposite: **Making music to encourage the nightingale to song.**
Engraving in Olina 1622, fol. 2v

➻+ CONTINUED

find themselves in a place where they hear the HUBBUB of several persons, they will speak; and almost *sparring* and contending and wishing to overcome, they make an extraordinary CLAMOUR with their *squawking*; but in addition to these manners Animals are also considerably enticed by means of scents. Cats, with CALAMINT sewn into a bundle after the manner of a *Ball*, will turn around and about it and run, from what is written of this, quite mad. Dogs, with the scent of their *Owner's shirt*, sometimes sleeping on it, become accustomed in this way to follow them as if under an enchantment. WOLVES and Foxes and *Buzzards*, with the scent of any Carrion, will gather to contend for it from many miles off. DOVES are so very lost to the scent of Cumin that public *prohibitions* on its use have been needed.

Our Nightingale, just as it exceeds other Birds in song, is also exquisite in its ACUTENESS and *sagacity* in picking out scents, whence it is seen that in the Countryside it readily TARRIES where there are scented herbs, delighting in them, as some have written, and particularly in one, that for its *pleasantness* is called Musk moss. For want of the same, this has even been attempted with TRUE MOSS: placing a grain or two of this, wrapped in a little Cotton-wool, on the canes that serve as *perches* in the Cage, stimulate it to sing; which having been successful, the very same has been tried with *equal success* in the Countryside with the Woodland Birds, with an UNGUENT composed of similar things.

So it must be, that pleasant and PENETRATING scents will heat it and *stimulate it* to sing; however you must leave off doing this *experiment* (I speak here of the caged Birds) when they are in their AMOURS, because having this moss or unguent continuously on the little *reeds and canes* harms them. The aforesaid little canes, when they are the ones upon which it PERCHES, must be in winter covered with *green bramble*.

Of the
STARLING

HERE IS NONE to whom the *Starling* (which in Latin is called *Sturnus*) is not known, it being seen in almost every Land in great ABUNDANCE. So its description could be *omitted* there, but so as not to interrupt the order adopted, we shall continue. It is a Bird of the size and garb of the BLACKBIRD, with the black colour that serves as a background on its whole body, *dotted* with light green which shimmers a little in both green and red (as can be seen in the colours of *Pigeons*) on the widest part of the wing, on the NECK, and NEAR THE EYES; the tips of the wings are dark greenish; the tail short and black; the beak *strong* and *longer* than that of the Blackbird; the feet reddish and the claws BLACK. The Female can be told from the Male because she does not display the same variety of colours that we have described above for the Male, also the Female has a little fleck in the WHITE OF THE EYE — in the Male it is *completely black*. The Starling chick in the nest is known in the same way, because it has its back, *wings* and tail black; whilst the head, neck and *stomach* are all grey.

It is seen in the MEADOWS, most of all in open farmland if there is *water and cattle*. It is also very often seen atop tall buildings, and on the rooftops and the DOVECOTES of houses, where it also nests in no other way but as the *Sparrows* do; it nests also in the countryside, in *great trees* — Chestnut in particular; and in woods and MOUNTAINS; two or three times a year, with four or five Chicks per *brood*. It is customary to catch those that are found on ROOFTOPS and buildings by placing on the walls some vessels of *unglazed terracotta*, made after the fashion of those *flasks* that countryfolk use, flat on one side, and the other bulging out, ➨

➡ CONTINUED

having on the *flat side* just so much of an opening that a hand may enter and ATTACHING these to the wall as shown in the illustration here; *Starlings and Sparrows* lay their eggs there with no disturbance whatsoever; when the birds are grown they are TAKEN OUT, returning divers times to nest there again; it is said that this is an invention of the *Flemings*.

As for their VICTUALS, although ordinarily they feed upon different *berries*, they will not scruple to lay waste to the Grape, to the Olive, and to *Fodder crops*, particularly Millet, Panicum and

Rosy starling
Watercolour and bodycolour
over black chalk, RCIN 927661

SORGHUM, and to almost every fruit, tearing into them very often with such fury that, owing both to their NUMBERS and to the ARDOUR and *violence* with which they proceed, when they arrive you may hear the air riven with a *horrible clamour* not unlike a hail-storm. They usually fly in TROOPS, availing themselves of this as a shield against the attack of the Hobby, *shrinking* down in the instant they are *attacked* into a ball in which, by lustily beating their wings, they create so much wind that they prevent the attacker from drawing near. TAKEN from the nest and caged, they serve as *Songbirds*, but more for learning to WHISTLE than their natural call. They may also be allowed to go about the house, becoming *marvellously* tame. ➤

Common or European starling
Watercolour and bodycolour
over black chalk, RCIN 927663

➤ CONTINUED

From NESTLINGS their food shall be *heart*, cut into little pieces of the size of a *writing-quill* and given them three or four at a time, some offered with a STICK until it is clear that they wish to eat for themselves, as has already been said; their feed is otherwise like that of the *Nightingale*. The woodland sort eats anything. It lives from five to six years.

Storno

Common or European starling (*Storno*)
A spherical nesting box can be seen in the background. Engraving in Olina 1622, fol. 17v

Of the
REDSTART

I N OUR TREE-LINED alleys and in *Copses* this Bird, that is called as above from the red colour it has, is found in the company of NIGHTINGALES or *Fig-peckers*. In Latin it is called *Ruticilla*, and PHENICURUS; the French call it the Nightingale of the Walls. It is in all respects similar to the Nightingale, a *little bigger*, and only different in its colouring. There are two sorts, that is to say GREATER and LESSER. The greater is in size a little smaller than a *Thrush*, it has its head somewhat flattened, or we would say depressed from where it rises up from the BEAK, which although it is quite broad at the base then narrows and grows *extraordinarily fine*; it is black, but not very dark; it has its head and neck ASH-COLOURED, with a few *splashes* of the colour of earth; the breast and stomach of rust-colour, with a few BLACK FEATHERS mixed with white, which outlining this colour make these parts appear as if wavy or *rippling*.

The flanks and the tail are similarly of BRIGHTER rust-colour, such as is seen on the breast of the *Robin*; the back and rump of *darker grey* than the head and neck, similarly outlined at the tips of the feathers with a little bit of RUST-COLOUR, but quite sparingly, and without *brightness*. The wings are the same. A little under the eye it is dotted as if with *rusty droplets*, that run towards the back of the neck; it has whitish legs, quite thin. *The Lesser* is identical in appearance to the Nightingale, rather a little smaller; it has its HEAD, neck and back of LEADEN COLOUR, that is ➡

⇥ CONTINUED

to say dark grey; the CHINSTRAP and on the breast are blackish, with some admixture of white feathers; where the *Stomach* begins it is dark ASH-COLOUR, and lower down towards the tail. Its head is the colour of rust; the wings *lighter* than the rump and tending almost towards *bay*; it has a very fine beak and feet and its mouth is black, both one and the other YELLOWING. It goes in the same places, at the self-same time, as the *Fig-pecker*; but it loves more the *mountain*, and fresh coolness, than the plain. It is seen in Summer, and the first two months of AUTUMN, going hence or retiring in November to flee the *harshness* of Winter. It sings in the Spring like the Nightingale. It nests in a HOLE IN A TREE, and sometimes in some dry twigs near the ground, or in a *crack* in some ancient ruin, laying two or three eggs. It moves its tail often, like the *Robin Red-breast*.

It feeds in the *Countryside* upon different Berries, most of all of those of the DOGWOOD and Figs or the *fruit of the Briar* as well as flies, *ants' eggs* and the like. Wishing to rear it in the house, for it to sing, it shall be given PASTA and HEART, and be tended with great care as it is more *finicky* and hard to please than the Nightingale; often it is also given crumbs of bread, and CHEWED WALNUTS. The male, which is chosen for its song, will have its breast more *spotted*, and be a more REDDISH colour. The woodland ones sing in the Spring, until the Summer comes in, ceasing to sing when it is has a BROOD. Its usual custom is to sing in the *morning early*, now upon the thicket, and now upon some ABANDONED BUILDING, being in this not very different from the *Solitary Sparrow*.

Those raised in the house sing at ANY HOUR of the day or night, and they learn to *whistle* and to *counterfeit* other Birds if ⇥

Common redstart
Watercolour and bodycolour
over black chalk, RCIN 927637

Common rock thrush
Watercolour and bodycolour
over black chalk, RCIN 927653

Opposite: **Common redstart**
(*Codirosso ordinario*)
with common rock thrush
(*Codirosso magire*). The
birds were illustrated together
in the *Uccelliera* as the greater
and lesser redtail but have
now been identified correctly
as different species. Engraving
in Olina 1622, fol. 46v

➤ CONTINUED

they are so taught. Of these two kinds, the *greater* one does better; it is more proper to this one than to the LESSER to stay inside, as has been said. They are taken with the *Spider's-web net*, and with the SMALL BOW. It lives from six to eight years.

Codiroʃʃo. ordinario.

Codiroʃʃo Magire.

Of the
THRUSH

HE THRUSH BEING good for *singing*, and for SERVING at Table, merits mention even though quite well known, and the most that is known of it should *already* be written down. Let us say, therefore, that in Latin it is called TUR-DUS, which is the generic name, being divided into *three kinds*, known by the additions to this name. Of these, the first is the GREATER, and is called *Viscivorus*, because for the most part it is seen upon trees where there is Mistletoe; it is a GREAT LOVER of these berries. In Italian it is commonly called *Tordela* (that is, Mistle Thrush), and in Lombardy it is called DRESSA or Missel-bird, and also *Storm-cock*; it is a little less big than the JAY, and darker than the others, and has *speckles* on the breast with a little bit of yellow tending towards RUST-COLOUR, not dissimilar to the female Blackbird.

In this matter of the *Mistletoe*, I should not neglect to say that *scaligero*, and after him *Aldrovando*, both DISAGREE with the common belief that this plant is *born* out of the Droppings of this Bird, and with this *good argument*: that if this were so MISTLETOE plants would be seen on many trees where they are not. Rather it is generated out of a *vital excrescence* of the tree itself, where it grows (almost as if from seed) as in the same way we see happen with OAK GALLS, and this is enough.

The second is the *Fieldfare*, that we call middle THRUSH, in Latin called *Pilaris*, which is the ordinary Thrush, although in Tuscany it is called *Bottaccio* (that is, THROSTLE). This is of that species that others call our own native Thrush, because it stays in our regions, being found in *Summer* in the *Mountains* and cool places; in Autumn, on the HILLS AND PLAINS; and in Winter in the *Maremma*, among the copses of Juniper and Myrtle. It is very delicate to eat, and the *Nestling* sings and whistles EXCELLENTLY well; we will not describe it for there is no danger that it is *unfamiliar*; we shall only give the advice on how to tell the Male from the Female; which is, that ➵

Song thrush
Watercolour and bodycolour
over black chalk; pen and ink,
RCIN 927683

➡ CONTINUED

the male has its breast quite SPECK-LED with black, and will have a larger *head*. To rear it and keep it, you should make use of the same rules as for the NIGHTINGALE.

The third kind is the minor, in Latin called *Illades* or ILIACUS, because it is notably speckled on the flanks and below the wings with *reddish* colour – this one is commonly called in Italy *Tordo Sassello* (or STONY THRUSH, that is, the Redwing), and it often appears escorting the others. Thrushes make their nest after the guise of the *Mud Swallow*, atop tall trees, and this is the reason why, as GREAT RAINS come in May and June, few of them appear in the Autumn,

Redwing
Watercolour and bodycolour
over black chalk, RCIN 927671

because the rain *spoils* their nests. They lay from four to five eggs. Their FOOD in the Countryside is *berries* and some GRUBS. In the house, if you wish to make them an *enclosure*, do so in the same way as for the Ortolan. They are taken with *snares* or with *small bows*, but in quantity with the Spider's-web net, that is, the fine BIRD NET or snare, and by Fowling or in *ambush*. It lives caged from five to six years.

It is to be noted that in EATING OF these Birds, as well as the *delicacy* of the flavour, there is also a particular benefit to the HEALTH, because they take quality from their feed; and just as the *Starling* is almost INFAMOUS for its repast of *Hemlock*; so these are for MYRTLE AND JUNIPER; their gizzards are helpful for the urine in some, to the *relaxation* of the Stomach in others.

Mistle thrush
Watercolour and bodycolour
over black chalk, RCIN 927685

Eurasian blackbird
Watercolour and bodycolour
over black chalk, RCIN 927679

Of the
BLACKBIRD

HE BLACKBIRD, IN Latin called *Merula*, is a Bird that has the greatest likeness and is most akin to the THRUSH, being of the same *appearance* in its shape, although different *colour*, and dwelling in the same places as the above-mentioned. Of these the male is all black, MULBERRY-BLACK, with the beak *yellow*, tending towards *Reddish*, and it has its legs equally yellow, but not so BRIGHT. The female is of the COLOUR OF SOOT, and has its throat and breast dotted with *dirty white*, and its beak is not so yellow. The little that it is yellow is more on the part *below* than that above. However, in addition to these colours, there are found by a QUIRK OF NATURE variants such as seeing them sometimes *patched* with white, and part white part black (which happens often), either owing to the QUALITY of the Land where they are born (such as those that there are born in *Norway*, which are all over white, it is thought because of the sight of the CONTINUAL SNOWS) or owing to the animal's own *nature*. It occurs among birds that there are many that change colour according to the *diversity* of the season, there being found some, most of all in the Autumn, that tend in colour from YELLOW TO BAY, that is to say to *chestnut*, and in that time they leave off *singing*.

It is found, as has been said, in the self-same places as the Thrush amidst the SCRUB and BRUSH and *Wooded places*, with Cypresses, *Juniper* and the like, in the Summer enjoying the freshness of the MOUNTAINS and other places, and in the Winter ✺›

➤ CONTINUED

enjoying the *Maremma*, being also at that same time in copses or groves in Gardens, and about INHABITED places. It broods twice a year, the first at the end of Winter, which but *rarely* comes out well, and the second in SUMMER, which succeeds *felicitously*: it lays from three to five eggs, which are all *speckled* with splashes of colours between GREEN and RUST-COLOUR; it makes its nest in *thickets*, or in some very thick and bushy *Sapling*, forming it of earth, hair and strands of dry grass, with a lining of softer material. It Sings on a par with the *Thrush*, and learns readily, being taught divers songs with a whistle, or the sound of the TRUMPET or *Drum* and the like: there are also those who accustom them to a few words. It Lives in the Countryside upon various *Berries*, and upon some fruit, also upon GRUBS or Maggots and *Grasshoppers*.

Wishing to make use of them for their song, you must have them from the nest, giving them for their food *Heart*, Meat, SOAKED BREAD and fruits. They are caught in the same manner as the *Thrush*. It is said that POMEGRANATE SEEDS will kill them. It must not be kept in small Aviaries, *persecuting* and *harassing* and giving trouble to the other Birds. It lives from six to eight years.

Eurasian blackbird (identified in inscription as 'white blackbird')
Watercolour and bodycolour over black chalk, RCIN 927689

Of the
MOUNTAIN
SPARROW

OR *Highland Sparrow*

HE DERIVATION OF the name of this *Bird* is taken from the place where it is used to be, which is the MOUNTAIN, whence it is called the *Mountain* or *Highland Sparrow*... This one stays ordinarily not only upon the sides of the MOUNTAINS, and HILLS, but also very often on the plains, betaking itself where it sees it can find something to eat.

It goes most often in the mud, and its flight is like that of ordinary *Sparrows*. It nests in the trees, and in some thicket, or fissure in the Mountainside. It lives upon *seeds*, and DIVERS GRAINS, and some worms or MAGGOTS; it also gives chase to *Flies*, and small *Butterflies*. Some are taken with spread nets, sometimes with the fine bird net or snare [like a spider's-web] or also with the LURE. It lives from four to five years.

Eurasian rock sparrow
Watercolour and bodycolour
over black chalk, RCIN 927650

Of the
TREE
SPARROW

ECAUSE IN THE form of its *body* this Bird greatly resembles the SPARROW, even though in its colours it differs not a little, it has come to be called *Sparrow*, and to distinguish it from the ordinary sort it is also called *Fuddled*, Foolish or FRANTIC, which it is thought to be in respect of its never staying still. In Latin it is called *Passer Stultus*.

It is of the SIZE AND SHAPE, as has been said, of the *Sparrow*. Its principal colour runs from yellow to UMBER earth, with patches throughout of the colour of *rust*; the beak goes toward red, and is STOUT AND SHORT; the tail and wings tend towards *black*, but in these the tips of the smallest feathers are white; the feet and legs tend from *yellow* to red.

Like the Lark, it is very often to be found near HIGHWAYS, but when it sees the *Traveller* not far away it takes flight, CIRCLING around and going a *little way* farther off: while it is perched, it bobs about CONTINU-OUSLY, raising and lowering its tail, making a gesture in its gait almost like that of the *Shrike* or lesser *Butcherbird*. It nests among TWIGS, where they are most thick and *bushy*, and sometimes in some hole in a *riverbank* or ditch, or in the SHELTER of some piece of earth, as has been said of the Lark; laying from four to five eggs.

In its way of living it is not very different from the GOLDFINCHES, because this one too feeds upon *divers seeds*, among others those of the THIS-TLE, perched upon which it is often to be seen. They are taken with the *Clap-net*, after which, when caged, they may be tended with *panicum grass*, millet, HEMPSEED or *Canary-seed*. It sings a little, but not particularly pleasingly.

It is never seen upon TALL TREES. It lives from five to six years.

Eurasian tree sparrow
Watercolour and bodycolour
over black chalk, RCIN 927615

Of the
SOLITARY
SPARROW

HIS BIRD IS counted by the *Writers* of *Natural History* among the BLACKBIRDS; in Latin it is called *Passer solitarius*, and in its traits and features it resembles the STARLING in size, with its beak somewhat longer, and a *little hooked* at the tip; its head, like its body, tends to be somewhat finer than otherwise, and FLAT *on* TOP; it is all *black* in colour, save that in the neck, and in the thick of the wings, where it has a hint of SHIMMERING colour between dark Turkish-blue (that we call *Turquoise*), and *Purple*, which is also sparsely scattered above the black. In the rest of the body and back, it has tiny patches of GREENISH OFF-WHITE. The Female is all dark, without purple, and with several little *patches* of dirty Yellow, as is seen in the female Blackbirds. It is found ordinarily among ANCIENT RUINS or upon the rooftops of great ancient Churches, where it makes its nest and sings very *sweetly*, being seen always alone. It sings for the most part in the morning. The nestling mostly learns how to *whistle exquisitely* whatever you like, whether ORDINARY words or little songs, and has also its own natural call which is very *fine*.

If you wish to *rear* them, it is necessary that they be well fledged; when they shall be HAND-FED with minced heart, eight or ten times a day; taking care in the *morning* for the first two ➛

Blue rock thrush
Watercolour and bodycolour
over black chalk, RCIN 927677

➡ CONTINUED

hours that you are up to feed it MORE GENEROUSLY, as it may have *gone hungry* in the night, giving it of the said heart three or four pieces of the size of a WRITING QUILL. Once it can eat on its own, its food shall be the *same* as that of the NIGHTINGALE.

To catch it, you should *observe* the places where it is found, and thither carry a caged bird, putting BIRD-LIMED twigs about the Cage, because *seeing* this other it will at once run at it to peck it, and will be caught fast; or in its *stead* you may put in the same place a LITTLE OWL with four limed twigs, arranged in the right way. Once caught, bind up its *wings* as has been said of the Nightingale, putting it into a Cage WRAPPED WITH PAPER, putting heart, and pasta, into the manger, hand-feeding it two, or three times a day until it EATS on its own. When you take away the *paper* do so a little at a time, so that it may not be disturbed.

They are particularly prized in *Genoa* and MILAN. They live, well kept, from eight to ten years.

Blue rock thrush (female)
Watercolour and bodycolour
over black chalk, RCIN 927662

Of *our*
NATIVE
SPARROW

HE SPARROW, THAT is our native one, unlike many other birds that are called by this same name, such as the *Solitary Sparrow*, or Canary or tree sparrow, is called in Latin *Passer*. Its traits and features are known to all, as it is to be found everywhere ... Sparrows (setting aside the species different from our own) are of two sorts: one of the Home, that in Latin is called *Passer Domesticus*, and the other of the Countryside, that is called *Passer Silvestris* or Campestris. The first lives in populated places and houses, beneath the roofs of Dovecotes, and in *cracks* in walls; they nest steadfastly a pair of times each year in the same place, having often seven or eight chicks, never fewer than four. Many place about the walls the *earthenware* pots that were described in the chapter upon the starling, about which a curious thing is the observation made by some *Hollanders* (who have an abundance of both one and the other kind), that if they put up the said vessels, some of ordinary terracotta, and some glazed in black, the Starlings go to the black ones as being of their *livery* and *plumage*, and the Sparrows to the ordinary ones, without ever being mistaken.

The second kind lives in the *countryside*, being in the day upon the plains and elsewhere where they may PECK, retiring afterwards to thickets and copses or to *dense woodland*. These are of lighter colour than the domestic, and have a beak tending more towards REDDISH. They nest in the trees, and in some thickets, and in the CREVICES of the mountains, making their nest of *feathers* and hay. They feed

Italian sparrow
Watercolour and bodycolour
over black chalk, RCIN 927628

⇛ CONTINUED

not only upon corn, and every other sort of grain (to which they do no little DAMAGE, going about in bands and great numbers) but also upon flies, *butterflies*, and the like, eating also without harm the seeds of HENBANE or Nightshade. They are most *sagacious* and *shrewd*, so much that they know the nets and BIRDLIME and Crossbows better than any other bird. They dearly love their own kind, so that when one has found enough to peck, it sings at once to call its *companions* thither, just as in the story told of PHILOSTRATE.

They are taken with the spider's-web net; driving them also from their holes with the *Weasel*, and they are HUNTED with the Merlin, that is the *Stone-Falcon*, or the Shrike already mentioned.

For the table they are NOT RECOMMENDED save only the Sparrow chick, the others being both tough and *bitter* and too hot a food due to their SALACITY, wherefore they were by the ancients consecrated to *Venus*: in Jewish law, Sparrows served for the sacrifice that was made by those cured of LEPROSY. The egg and the *brain* of these birds are used in Electuaries for *satyrism*, for husbands who are cold and have little VIGOUR.

Italian sparrow
Watercolour and bodycolour
over black chalk, RCIN 927649

Of the
GREAT OWL
and the
LITTLE OWL
AND THE MANNER OF *fowling with these*

THE GREAT OWL, which is sometimes called the *Barn Owl,* is that great nocturnal Bird similar in form to the LITTLE OWL but as big as a *Hen*, with feathers on the side of the head that appear like two little HORNS, yellow in colour mixed with outlines of black, and which in Latin is called *Bubo*. With this fowlers go birding for large animals like CHOUGHS, Crows and *Kites*. With the Little Owl they go birding for little Birds of every kind.

The GREAT OWL dwells in grottoes and holes in Trees, and in ancient ruins or *cracks in walls* and in the the roofs of ABANDONED houses. It nests upon cliffs and in lonely hidden places, and a *curious thing* known of its nesting and BROODING is this, that in hatching it is out of the ordinary, coming out first with the tail, rather than the head. It is extremely well *armed with talons* as well as the strength and SIZE OF ITS BEAK, whence it hunts for itself even at night, *preying* upon different birds, and when it is attacked it DEFENDS itself bravely. It eats various things, but above all it seizes upon *meat*.

The manner of FOWLING with it is this. In a place where the aforesaid *Birds* are seen to pass by, select a tree not too close to any others. If too full of TWIGS and BRANCHES, cut off as many as seem expedient, coating the remainder thickly with *birdlime*. Or else *lime-twigs* may be tied to it, whichever may suit the FOWLER better. Not very far off from this you shall set the *Great Owl*. When the Birds see that it is there, they will run at it all together as if QUITE MAD, and ➥

※→ CONTINUED

---

after turning about it several times, and plucking out some FEATHERS, they will dash themselves against the tree, and remain caught fast. A great number of Choughs and *Crows* are taken in this way, and so also are some KITES, of which they make another very beautiful Hunt with the same Great Owl and a *Falcon*, known as the sacred or SAKER FALCON. The *Great Owl* is carried by the *Falconer*, and where they see Kites it is let fly, with a FOX'S TAIL attached to its feet. The Great Owl flies *close to the ground*, nor does it go very far, and alighting likewise upon it, the Kite comes STRAIGHTWAY down close by; then the Falcon is set upon it, and now the Kite leaves off the *curious prank* of the Great Owl and flies up, and it goes FENCING and PARRYING with its twists and turns as best it can the *onslaught* of its enemy, to the great diversion and *entertainment* of whoever may see it.

The LITTLE OWL is known to all, and so no more will be said of its *characteristics*, but on the contrary we will speak of its nature, and how it is HUNTED. It dwells for the most part in places where the *air is thick* and on the plains; those few that dwell on the MOUNTAIN are different from the ordinary sort in their legs, and feet, which are *feathered*. They are seen at the breaking and at the closing of the day, at which time they go seeking and procuring their VICTUALS, which consist of Mice or *Rats*, Lizards and FROGS. They nest in the two last months of the Winter. If being reared they should be tended with *meat*, being kept and maintained thereafter with the same. It is a property of the Little Owl to move about in ATTITUDES and GESTURES, raising and lowering itself, gazing sometimes very *rapt* and *fixedly*, and sometimes turning its head hither and THITHER, whence comes the quip remarked against some *women*, calling them Little Owl. They are particularly disliked by the WREN, the Crow and the *Jay*. It serves for Hunting in several ways, now with nets, now with lime-twigs; with the CLAPNET, when seeking any Bird you lack, use it to serve as a lure, as was said in the *Chapter on the* DANCING-BIRD or Wagtail, and with birdlime as with the *Great Owl* above, with which an infinite variety of Little Birds are taken. This is done toward the end of *Autumn*, and in Winter. It serves also for the AMBUSH, and for many other occasions. It lives from eight to nine years.

Eurasian eagle owl
Late sixteenth-century Italian
artist. Watercolour with touches
of bodycolour over black chalk,
RCIN 928743

Of the
FISHER
BIRD

Common kingfisher
Watercolour and bodycolour
over black chalk, RCIN 927655

HIS HAS DIVERS names, but most, recalling how it takes *Fish*, calling it FISHER and Kingfisher: in Rome and in Tuscany it is called the *Bird of Saint Mary* or 'of the Madonna', from the large amount of blue seen on it, like that which the Painters use in their PAINTINGS of the Madonna. In Lombardy it is called by many the *Merlo Acquarolo*, or DIPPER; and by others Piombino, that is the Plummet or *Plumb-bob*; in Latin it is called ISPIDA, and it is thought to be a kind of *Halcyon*.

It has a beak half as big as its whole size or a little less, deep black and *very sharp*; the head is covered in little feathers of LIGHT TURQUOISE or Turkish-blue, that at their tips appear edged with a *little white*; the wings are adorned in the same way as the head, but in a BRIGHTER BLUE... the rest of the back is blue, with admixture of a little green, that tends towards *Sea-water* or AQUAMARINE; the tail is of the same; the eye is placed in the middle of a patch of the colour of Rust between the *beak* and the *wing*, beneath which there is another turquoise or TURKISH-BLUE, that extends from the junction of the beak beneath the eye as far as the BEGINNING of the breast, which is all of the same colour of Rust... it has quite *short* and *slender* legs, and they are red in colour.

Kingfishers are found throughout ITALY, along rivers or close to springs, *perching* by the banks upon some tree or rock that stands out so that, ESPYING THEIR PREY from there, they might the more easily dive down upon it in time and thus obtain it. They live upon little *fish*, GRUBS and other *low creatures* that are found in these waters; in winter they are seen also by springs near populated places, most of all in *times of ice* and GREAT COLD...

It nests by the Banks in some pit dug in the *stone* or a *hollow* that it may find there, making its nest of ears of wild reeds, with four or five Chicks at the least. It flies in RARE FASHION, skimming the water. When they fly they *shriek* in such wise that they are heard uncommonly far off. Many keep them, DEAD AND DRIED, affixed in their *Chambers* for their beauty, and some make of them a master of the *storehouse*, being of the opinion that they protect the stuff therein from becoming *wormy* or MOTH-EATEN. Others say that these birds change their feathers every year, which is *false*. It lives from four to five years.

HE RESEMBLANCE BETWEEN divers little birds of *iridescent* shimmering green and yellow, divided into light and dark, has led to difficulty in DISTINGUISHING the same; and many *confuse* the Siskin, or *Barley-bird*, that in Latin is called Ligurinus, *Spinus* and ACHANTIS, with the Serin that some call in Latin Citrinella or THRAUPIS; however, to any who are properly observant the differences are very apparent, as described fully in this chapter, and in that on the *Siskin*.

The SERIN is a small bird; lively, cheerful, with a short beak, quite rounded, with the throat, *breast* and stomach of yellow tending towards GREEN. Close to the wings it is studded with dark green, mixed with colour like *umber earth*; on the head it is speckled in the SAME MANNER as on the breast, with the cheeks and back *splashed* with *patches* of light and dark of the said two colours; the tips of the wings being darker than the rest, and ALMOST BLACK. The rump is of the colour of the breast, and even lighter: the tail the same as the TIPS OF THE WINGS, and somewhat divided, almost in the manner of the *Swallow*.

Its Song is DELIGHTFUL in concert with other birds, but alone, having a call that is *quite short*, and repeating CONTINUOUSLY the same, it is not altogether satisfying. Its manner of *singing* has gained it the name among the French and in PIEDMONT OF TARIN, alluding to its *well-worn* song... It nests in our countryside like a Native bird, and in our *gardens* in thick, dense trees and most of all in CYPRESSES, building nests of wool, hair and feathers, and laying *four or five* eggs per clutch. If you wish to rear them, they must be taken when well FLEDGED, and be kept in their own nest, for want of which a *false* one should be made of WOOL or HAY. Their food shall be like that of the Goldfinch, and when grown they should be given *Hempseed*, or Panicum.

The COUNTRYFOLK take them with the clap-net or with *lime-twigs*, as has been said of the Goldfinch, and quantities of them are taken, as they are very SIMPLE-MINDED Birds, so that if one of them flies down, the *whole flock* does the same, and even though they ESCAPE the nets once they still return to them. The time to take them is the *Autumn*. They live from four to five years.

Of the
SERIN

European serin
Watercolour and bodycolour
over black chalk, RCIN 927601

Of the
CALANDRA
LARK

IF ANY BIRD deserves to be *prized and esteemed* it is this one, because in this one alone can be found all the QUALITIES otherwise found separately in many. The *Calandra Lark* is a sort of Lark, but a little bigger, so that some call it the GREAT LARK; its name in Latin is the same, only with a little more aspiration: *Chalandra*. In the vulgar tongue it is thought that its name alludes to CALARE, that is to *lower*, or *diminish*, which is what it does in its call: always starting high up, and LUSTILY, but then always growing less and falling away.

Its build is NOT DISSIMILAR from that of our own *native Lark*, although it is bigger – in its proportions, quite like the THRUSH. Its foreparts are light greenish with some *splashes of black* or dark grey on the breast as the Thrush has; behind the wings and tail it is the colour of UMBER EARTH. Two fingers'-breadth below the beak it has a ring of black feathers, like a *necklace*; its head is WIDER than the Thrush's, and its beak is *shorter* and *bigger*; while its legs are like other Larks'. The male is STOUTER than the female, and has more *black* about the neck.

The woodland variety sings like the other *Larks*, but with more varied calls. In the FIRST YEAR that it is caged it does not do much, and because it is a *strong bird*, and remembers the COUNTRYSIDE, it can be a little wild; wherefore you must ➺

Calandra lark
Watercolour and bodycolour
over black chalk, RCIN 927664

➥ CONTINUED

---

either bind its wings, or clad the upper part of the *cage* with a piece
of cloth drawn GOOD AND TAUT, so that in dashing itself against
it and striking it with its head it will not *kill* or wound itself (this is
EXCELLENT ADVICE, not only for this Bird but also for many
others). To keep it as a songbird it must be had either as a *Nestling*,
trying whenever possible to take one from an AUGUST clutch, or
young enough to have its first *moult of feathers* in the cage. As well
as their own natural call they learn MARVELLOUSLY well those
of others, most of all the Goldfinch, LINNET, Swallow, *Canary*
and the like; it can also learn to counterfeit Chicks, *Hobbies*, Cats
and others. In rearing them you should feed them minced HEART
AND PASTA, Spelt, Oats, *mixed Birdfood*, and fresh crumbs of
bread, not neglecting to keep continuously in the cage a little piece
of CUTTLEBONE. Whenever you wish that it should learn a spe-
cific new call, it must be kept in a place where it does not hear
other bird songs or CALLS that it might *learn*.

They nest on hard ground and among SOWN FIELDS, making their nest like other Larks in the shelter of some *Clump* or
CLOD OF EARTH, well covered with grass, with four or five eggs.
They are caught like other Larks with Wall-nets, the FOWLERS
being in a little *Hide* or cover made of *leafy branches*; the spreading
of the nets is done for the most part in a place near to water, where
the birds go to DRINK; equally they may be *taken at night* with a
long-handled net and a *light*.

Greater short-toed lark
Watercolour and bodycolour
over black chalk, RCIN 927620

HAT WHICH WAS said in the description of our own *native Lark* serves in many aspects with regard to the present one as well, so it is more TROUBLE than it is worth for me to *expand* upon it anew, particularly as its *traits and features* are identical, save in a little tuft that differentiates it from the other and which springs up within the bourn of its TWO EYES, extending over the head. This is black in colour but not very *dark*, standing out a little with four or six bigger feathers from the other *feathers* like a little crest, whence it is called CRESTED LARK, and in Latin *Galerita*. Its body is a little bit whiter than the others', to which it is also judged *inferior* in song. The Male of these has its breast quite PATCHED with black, with its *beak* and *head* bigger. It flies differently from the other Larks, almost always alone, and not keeping STEADY or STABLE but going now high, now low, accordingly as it is borne by the Wind or the *freshness* of the air. It stays for the most part upon high tops in the fields, or upon the banks of DITCHES, and on the *highways*, where from the dung that it finds it procures its living, in particular in the WINTER.

As regards its nest, and how it broods, it is as above; though this one nests nearer to ROADS than the other. Whosoever wishes to raise them from the nest shall *observe*, as has been said with regard to taking them, that they must be well fledged; and will HAND-FEED them with patience, with *minced heart*, opening their mouth gently, and taking care when they put some in, not to envelop or cover the *tongue* with the MORSEL and force it upon them, for they would suffer. Let the *little pieces* be of the size written, and a little long. It is written by proven AUTHORS that the same birds reduced to a *powder*, and this being mixed with appropriate liquor and *drunk*, or eaten, and most of all BOILED for four days continuously, frees one from the pains of *colic*. The words of MARCEL VIRGILIO upon this matter are the following:

'Put the lark, with its FEATHERS, in a vase of clay covered in lime plaster and put in the oven to be *burnt*, so that it can be reduced to POWDER. Then the very fine ground powder is collected, and two or three spoons of that mixed with *hot water* are to be taken for three or four days. This is an INCREDIBLE remedy for colics, certainly useful, as it justly seems to be better than any other *medicine*.' ➤

Of *the* CRESTED LARK

Crested lark
Watercolour and bodycolour
over black chalk, RCIN 927641

➺ CONTINUED

The manner of MAKING THIS POWDER, *Pliny* describes it with these precise words:

'Then the ASHES OF BIRDS or of other animals has to be made as *indicated*. The bird, or any other animal that you want to be burnt, is put in a NEW JAR, then place a lid upon it and spread it over with clay and put it in a *hot oven*, with a very small breathing-hole, to be BURNT.'

Others would have it that this should be done as has been said by BOILING, reducing it after the manner of a *consommé*; which having the virtue of *dissolving* benignly, for this reason it is reputed as suitable for this ILLNESS. Others write to the same effect, that the Lark having been slaughtered, its heart must be taken out while still warm, and then dried, sewn to a RIBBON OF SILK, and worn against the bare flesh tied to the left *flank*.

Della Porta, who is a scholar of the SIGNS and MARKINGS of things (so exalted and praised by the modern *German Doctors*) justifies this secret in these words:

'Talkative animals are not EFFICACIOUS for relieveing the suffering of the *colic*, no more than are talkative people: too much talking, indeed, releases a breeze from which often the ILLNESS arises; the birds' quality, therefore, is impressed upon us in the way they *remove* the illness.'

Which, if it be true, it would perhaps merit the COMPASSION of all if, in order to escape such a great ill, they were to cling fast to such an easy recipe.

Of our own
NATIVE LARK

THIS NAME OF Lark, in *Latin* called in general *Alauda*, includes divers sorts of the same Bird which are then distinguished by the addition of CRESTED, Greater, *Lesser* and the like: whence in saying simply 'Lark', it is understood to be the simple or more commonly the non-crested one, which the FOWLERS sell by the dozen as a bird for *eating*, on which we reflect in the present Chapter. Because it is quite well known, there is no need to make too EXACT a description of it, nevertheless it will be as well to say that it is *not much bigger* than a Sparrow, and a good deal LONGER.

It is of the colour of earth; on the *breast* somewhat lighter and tending towards the colour of *cinders*... it has white legs, and in these it has the last TALON quite a lot longer than the others. It nests on flat ground on *firm earth* in the shelter of some clod or mass of earth, and lower rather than higher. It makes its nest of the FIBRES of plants and *dry grass*, with four or five eggs, nesting three times a year, that is in May, July and *August*: it rears its Chicks very SPEEDILY, finishing its brooding in a FORTNIGHT AT MOST, and rearing in much less, wherefore it is useful for any who wish to take them to be advised to do so as soon as they are *well-fledged*, for by leaving them any longer you run the risk that they may ESCAPE. In rearing them they shall be tended like the Nightingale with *chopped heart*, but once they are reared so as to cost less they may be given HULLED WHEAT, Spelt, Oats, mixed Birdfood and Millet. The Male has the *talon* mentioned above so long that it goes past the knee; and it has two BLACK PATCHES on the neck, one on either side; on the breast it is darker, *greyish black*, and it is stouter. ➤

➤ CONTINUED

Its Song is DELIGHTFUL because it is so varied, full of *trills* and the *warbling* that is called roulade, and diverse *diminuendos*; it sings ordinarily in the morning in a clear SKY, rarely on the ground. In its flight it turns continually, climbing, and *singing*, taking pleasure FROM TIME TO TIME with a finely tuned motion of the wing to *maintain* itself in the air, whence going down little by little at the last it descends with such FURY that it is more diving than *descending*.

The Countryfolk take them in great quantities in the *Autumn* right up to ALL SAINTS' DAY with the Clap-net, situating the nets near to some sown field, pasture or meadow or little *knoll*, where the birds may pass, with two Larks to serve as LURES, in such wise that it is possible to raise up now one, now the other; putting also in the area between the *two nets* some of these same Larks, dried, to serve as DECOYS. In addition you will also need a bird-call whistle, with which you will imitate their *Piping*, taking particular care to COUNTERFEIT it well and repeating it several times as soon as you see that they wish to come down. They are also taken at any time that they are found at night, with the *Long-handled net*, the CATTLE-BELL and the Dark-lantern, and this practice is called Low-belling. It lives from eight to ten years.

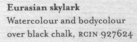

Eurasian skylark
Watercolour and bodycolour
over black chalk, RCIN 927624

Of the
WOODLARK
and
MEADOW LARK

S WELL AS the Larks described *above*, there are THREE other sorts: the present one, which is known only in the *Countryside* of ROME and which is placed among the Larks that are not *tufted*, whence it falls under the Latin title of ALAUDA NON CRISTATA; then there is the Meadow Lark, which in ROME the Fowlers call *Calandrino* or Little Calandra Lark, of which we will speak in this Chapter; then there is the third, the CALANDRA LARK or *Great Lark* which is described in its separate Chapter.

The WOODLARK is no bigger than it is shown in the adjoining image, which is quite *diligently* done. As regards its COLOURS, it is similar to the tufted Lark, but just a little smaller, and it has on its head a *row of feathers* that are a LITTLE DARK, and which in their shape resemble a coronet. Its breast is *whitish* or off-white, sprinkled with DROPLETS of *dark grey* or *black*; the head too, near the eyes and beside the beak. The CHINSTRAP below the throat is off-white, but darker towards the breast; the *neck*, the rump, the *wings* and tail tend towards the colour of BAY or faded *chestnut*.

This bird sings quite FINELY, not only by day but also by night, like the *Nightingale*, to achieve this it must be reared as a nestling, as described above; once REARED, it should be fed Panicum and *Millet*. In the female the CORONET is not as dark as in the male, and the *back claw* or *spur* in the male is so long that it goes past the knee. It nests in broad valleys, where the trees are dense and bushy, making its nest after the FASHION of our own common or *ordinary Lark*.

The Woodlark is good for SINGING, provided that you succeed in

Woodlark
Watercolour and bodycolour
over black chalk, RCIN 927625

putting it at ease. It is taken like the other *Larks*. It lives from eight to ten years.

The *Meadow Lark*, or LITTLE CALANDRA, is finer and smaller than all the others, from which it also differs in being SPECKLED with *yellowish flecks*, being for the rest here black, and there *tawny*. The feathers of the tail grow white at their tips; those of the rump are dark; and its beak is quite long and delicate. It is usually seen on the ground, except when it fears the FALCON, to flee from which it retires among the *branches* of some nearby tree.

Among the SINGING Larks, praise is given to this one, because of which it is very HIGHLY PRIZED, as well as being a *rare bird*, and reared with difficulty; it too should be tended according to the rules given for the *Nightingale*. It lives from three to four years.

HE CONVENIENCE OBTAINED from *Birds* with regard to their feathers is EXT-REMELY great; considering in addition to the use invented by the *Indians* to make caps and garments of them, they are also used for the GAUNTLET CUFFS of gloves, the covers of *muffs*, for *stomachers*, and in many other occasions and MAN-NERS. Therefore it can only be good to give a sure method of *dressing* and keeping them without spoiling.

It will be OBSERVED then that we shall make use of Birds that have not died of *themselves*, but which have been SLAUGHTERED. Nor are they good every time because, as has already been said, *most of them* at a certain time of the year MOULT their feathers...

Let it also be *understood* that this work is to be done with the WANING MOON. The rule to be followed is this: *Spreading* out the feathers from the neck by BLOWING upon them, uncover just so much that you may, with a *little knife* that cuts well, make an opening; this you will continue over the JUNCTION of the WINGS, proceeding with the cut along the

flank, up to the *tip* of the *tail*. You will then pull with your FINGERS with patience, *flaying off* the flesh, and cutting *nerves* or whatever may hinder, so that it is detached, BREAKING it where it joins to the wings and *thighs*; such is this task. The head if it is small may be left, putting SLAKED LIME mixed with *Myrrh powder* into the beak to dry out what little flesh may be there, otherwise it shall be FLAYED off, pulling the skin inside out.

This way shall *serve* for those who wish to make a MODEL from it for any work, or to adorn any *study*. Having made a stuffing of COTTON WOOL in which there should be a little *Wormwood*, and sewing back up the cut and having arranged and fixed the *Wings* and Legs with COP PER WIRE, these will serve most GALLANTLY indeed. For making something else of these, such as for example desiring to have that *shimmering* Green that is on the head and the necks of Ducks to make covers for GLOVES or MUFFS, you shall go this other road.

Having *flayed* off the SKIN, stretch it out with the feathers on the *good side*, so that they shall not ➥

Method of dressing the SKINS OF BIRDS *for* DIVERSE USES

Green woodpecker
Watercolour and bodycolour
over black chalk, RCIN 927690

European bee-eater
Watercolour and bodycolour
over black chalk, RCIN 927674

➤ CONTINUED

be RUFFLED, upon a little wooden tablet, or the *bottom* of a box. Then stud it with FINE WIRE on every side so that it becomes *properly taut*, and having taken away from it whatever there may be of FAT or FLESH, and having mended it with silk if any *tear* is made in it, soak the said SKIN with a glue made of a *handful of flour*, a pinch of fine COMMON SALT, and as much good *white wine* as shall be sufficient to temper it and reduce it like a glue for *lime-coating*. Having BEDAUBED it equally all over, set it to dry out in the shade facing towards the *North*. If the skin, having been cleaned of this Glue (which comes away in LITTLE SCALES or FLAKES if you scratch it with a little knife), shall yet appear to retain some *little humidity*, return anew to plaster it as if with a POULTICE, and dry it out. Once *dried out*, as they will be, the skins should be put into a box, making a *bed* at the *bottom* of the above-mentioned WORMWOOD, or sawdust of rosewood. If you wish to give them a scent you may first, before taking them off the little *tablet*, being cleaned as they are of the

GLUE, give them a coat or two with a *sponge* of any *perfumed* composition as your whim may please you. Birds whose skins are suitable for this are Ducks, PHEASANTS and *Peacocks* for the Shimmering that they have on the neck.

For the *Swan*, Buzzard and STORK, owing to their being hot, instead of Wine in their curing and tanning *Vinegar* is used, in which there has been dissolved a little COMMON SALT and Alum, giving it more than *one covering* of this mixture according to need.

Of other Birds, for their *loveliness*, there will be taken RAZOR-BILLS, that are called Sea-crow or *Sea-magpie*; the Green Woodpecker; and the Bee-eater, which some call GRAVLO and some *Ghiovaro*, and others, for how it eats Bees, call the *Bee-Wolf*. The River Halcyon, that is the *Kingfisher* or PLUMMET, and the like: pieces of these, if not made into charming pictures as the *Indians* use them, may be put together in some DESIGN, and such things shall be made of them that will turn out very *Lovely* to look upon.

Index

References given in bold are to pages carrying the illustrations of
bird drawings in the Royal Collection, for which scientific names
of the species depicted are also given alongside the common names
indexed. See the notes for the reader on p. iv for more on the
source of these identifications.

Acknowledgements

Royal Collection Trust is grateful to the following individuals and colleagues for their assistance in putting together this publication: Rea Alexandratos, Polly Atkinson, Brian Clarke, Kate Clayton and James Stedman at First Edition Translations, Jacky Colliss Harvey, Kate Heard, Sarah Kane, Helen Macdonald, Ocky Murray, Vicki Robinson, Simon Toop, Debbie Wayment, Nick Finegold and the team at Zebra, and Eva Zielinska-Millar.

Song thrush (leucistic)
Watercolour and bodycolour
over black chalk, RCIN 927688

Published 2018 in the United States by Yale University Press
and in the United Kingdom by Royal Collection Trust
York House
St James's Palace
London SW1A 1BQ

Yale University Press books may be purchased in quantity for educational, business, or promotional use.
For information, please e-mail sales.press@yale.edu (U.S. office) or sales@yaleup.co.uk (U.K. office).

ISBN 978-0-300-23288-2

Library of Congress Control Number: 2017960052
A catalogue record for this book is available from the British Library.

The translation of this book has been funded by SEPS – SEGRETARIATO EUROPEO PER LE
PUBBLICAZIONI SCIENTIFICHE, via Val d'Aposa 7, 40123 Bologna, Italy (www.seps.it; seps@seps.it)

Translation from the original Italian by Kate Clayton in association with First Edition Cambridge, UK
Edited by Sarah Kane
Index by Vicki Robinson
Designed by Ocky Murray
Production management by Debbie Wayment
Typeset in Granjon and Mrs Eaves
Colour Reproduction by Zebra
Printed on Munken Lynx Rough 170gsm
Printed and bound in Italy by Graphicom srl

This paper meets the requirements of ANSI/NISO Z39.48-1992 (Permanence of Paper).
10 9 8 7 6 5 4 3 2 1

front cover: Common nightingale, RCIN 927607, and European robin, RCIN 927626
page v: Engraving of an aviary in Olina 1622, fol. 67v

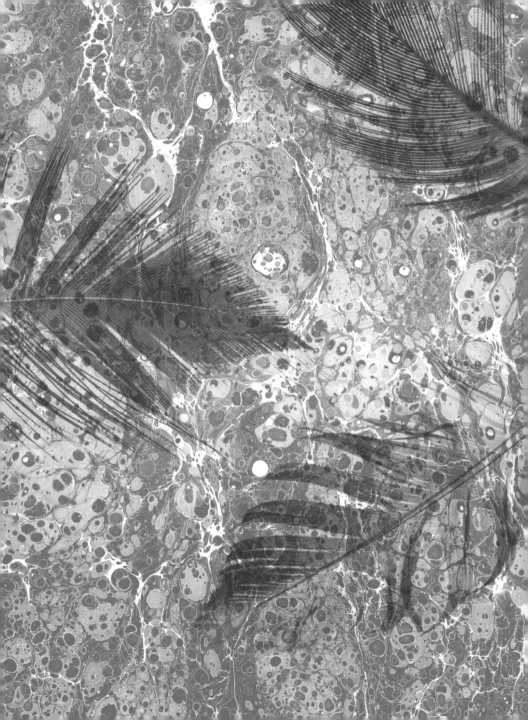